Energy Performance of Buildings

Authors

George Baird, Ph.D.
Michael R. Donn, M.Sc.
Frank Pool, M.E.
William D. S. Brander, M.Sc.
Chan Seong Aun, B.BSc., B. Arch.

Energy Research Group
School of Architecture
Victoria University
Wellington, New Zealand

CRC Press, Inc.
Boca Raton, Florida

Library of Congress Cataloging in Publication Data
Main entry under title:

Energy performance of buildings.

Bibliography: p.
Includes index.
1. Buildings--Energy conservation. I. Baird, George.
TJ163.5.B84E57   1983   690   83-7246
ISBN 0-8493-5186-3

This book represents information obtained from authentic and highly regarded sources. Reprinted material is quoted with permission, and sources are indicated. A wide variety of references are listed. Every reasonable effort has been made to give reliable data and information, but the author and the publisher cannot assume responsibility for the validity of all materials or for the consequences of their use.

Direct all inquiries to CRC Press, Inc., 2000 Corporate Blvd., N.W., Boca Raton, Florida, 33431.

© 1984 by CRC Press, Inc.

International Standard Book Number 0-8493-5186-3

Library of Congress Card Number 83-7246
Printed in the United States

PREFACE

This book deals with the concerns of everyone involved with the use of energy in buildings. It is written principally for those with a direct professional interest in the energy performance of buildings:

- Energy planners and those involved in the drafting of building standards at national and regional level.
- Building owners — whether occupiers, investors, or developers.
- Building designers of all persuasions — architects, engineers, energy consultants, and so on.
- Building users — principally the building manager, the energy manager, and the building operator.
- Energy researchers, educators, and students of the field of energy and buildings.

Much of the material in this book stems from the authors' work at the Energy Research Group of Victoria University of Wellington's School of Architecture. Material has also been drawn from the results of a few other systematic empirical studies of building energy use. The driving force behind all these studies has been a desire to understand the factors that affect energy use in practice, a need to find ways of measuring and evaluating building energy performance, and doubts about the accuracy of the design predictions provided by current building simulation methods.

We recognize that the reader's motives may be more pragmatic and will be mainly concerned with reducing the cost rather than the consumption of energy. However, after the most economic fuel has been selected and the lowest tariff negotiated, it is the amount of energy consumed and the maximum demand that determines its cost. Energy consumption figures also allow international comparisons of building energy performance to be made — cost data is much less meaningful in this context. The material presented here is designed to assist those concerned to achieve the best possible energy performance from their buildings, no matter where they are located.

We have arranged the material to allow those with particular interests to find the relevant sections of the book quickly. After a brief historical introduction to energy use in buildings we present a new framework for classifying the concerns of nations, owners, designers, and users of buildings. This is followed by chapters dealing, in detail, with the concerns of each of these groups — and each chapter is substantially self-contained. Thus, those with specific interests could, for example, study the introduction and the framework (Chapters 1 and 2, respectively) and then move directly to the chapter dealing with their specific concerns. The final chapter looks to the future of building energy performance studies. A perusal of the contents will clarify which chapter or chapters are most relevant for a particular reader. We strongly suggest, however, that readers should at the very least skim through the other chapters to gain an appreciation of the concerns of other groups.

In order to make the information relevant to as wide an audience as possible we have presented it in both SI (Système International) and IP (Inch-Pound) Units. Since one of our main aims is to give the reader a feel for the measurement of building energy performance we thought it essential to provide both.

The authors are all members of the Energy Research Group. The Group commenced its activities in 1977 with a modest pilot study of the commercial and institutional buildings in the central business district (CBD) of Wellington, the capital city of New Zealand. The Wellington CBD is now surveyed annually and members of the Group have completed systematic surveys of the 1000 commercial buildings located there. They have also carried out a major study of energy management in government buildings, investigated the energy embodied in light construction buildings, surveyed energy use in most of New Zealand's primary and secondary schools, and conducted a detailed study into energy end use and performance monitoring in commercial buildings.

The Group is now embarking on a major study of the opportunities for more efficient use of energy in commercial buildings, the outcome of which will be recommendations, for policy makers at the national level and for owners and managers at the individual building level, on the economics of a range of energy conservation measures.

For up-to-date information on the Group's activities, readers are invited to write directly to The Energy Research Group, School of Architecture, Victoria University, Private Bag, Wellington, New Zealand.

George Baird
Michael R. Donn
Frank Pool
William D. S. Brander
Chan Seong Aun

THE AUTHORS

The authors are all members or former members of the Energy Research Group at the School of Architecture, Victoria University of Wellington, New Zealand.

Since its inception in 1977, the Group has carried out contract research projects in the field of building energy performance for five national organizations — the New Zealand Energy Research and Development Committee, the Ministry of Energy, the Department of Education, the Ministry of Works and Development, and the Building Research Association of New Zealand. Members of the Group have authored ten major contract research reports and are in continual demand as advisors and speakers on all aspects of energy and building. In the course of the last 6 years (1978 to 1983) the results of the Group's work have been extensively published, with more than 20 papers in New Zealand scientific and technical journals and a similar number in the proceedings of major international conferences.

The Group is currently investigating energy conservation and energy management in commercial and institutional buildings, developing energy prediction methods for commercial and domestic buildings, and studying means of technology transfer in these fields. A further book entitled *Energy Management of Buildings* is in the course of preparation.

George Baird, Ph.D., is Reader in Building Environmental Science and Director of the Energy Research Group at the School of Architecture, Victoria University of Wellington, New Zealand.

He received his engineering education at Glasgow and Strathclyde Universities in Scotland and at the National College for Heating Ventilating, Refrigeration, and Fan Engineering, London. Following that, he spent 6 years with the Building Services Research Unit at Glasgow University, investigating air transfer and its control in a full-scale experimental hospital ward and carrying out studies of the design and performance of disposal chutes. He was awarded the degree of Ph.D. for his research during this period.

The next 6 years (1969 to 1975) were spent at the Scott Sutherland School of Architecture, Aberdeen, Scotland, as a Lecturer in building science and services. At the end of that period he was asked to assist in the setting up of courses and facilities for the new School of Architecture that was being founded at Victoria University of Wellington. His major efforts have been directed towards the organization of courses in building environmental science and services, and the establishment of a research program into the energy performance of buildings.

Dr. Baird has been director and principal investigator for over ten major research projects, has been responsible for the writing of the final reports for each, and is sole or principal author of some 30 technical papers. He is active in professional circles, being a Member of the Institution of Professional Engineers, New Zealand; the Chartered Institute of Building Services, U.K.; and the American Society of Heating, Refrigerating, and Air-Conditioning Engineers; and is a member of numerous committees concerned with such matters as district heating, building energy standards, energy conservation, and building research.

Michael R. Donn, M.Sc., is a Lecturer in the School of Architecture at Victoria University of Wellington. He has been involved in building environmental science research, teaching, and consulting for the past 7 years, taking a special interest in both building energy use and architectural aerodynamics.

Mr. Donn received his M.Sc. degree in Physics from Victoria University in 1976. His research activities include directing contract research projects examining energy use in schools and commercial buildings, plus developing a low energy design guide for New Zealand domestic building. In addition, he coordinates the New Zealand contribution to International Energy Agency project "Passive and Hybrid Solar Low Energy Buildings".

Other activities in which Mr. Donn is involved are research projects examining the energy use consequences of design decisions with special reference to the interface between energy or building systems and their users, exploring the application of scientific information in design, and considering the practical application of energy conservation measures. During his involvement with the Energy Research Group he has authorized 15 technical papers and research reports, and co-authored a further 7 on building energy performance and related topics.

Frank Pool, M.E., is an Energy Research Fellow with the Energy Research Group at the School of Architecture, Victoria University of Wellington. He received his B.E. and M.E. degrees in Mechanical Engineering from the University of Canterbury, New Zealand, where he specialized in Energy Studies and worked on the computer prediction of commercial building energy use. He has been involved in energy-related research since 1977, first into internal combustion engine design, then with the New Zealand Electricity Division dealing with economic studies, and subsequently with the Energy Research Group at the School of Architecture, Victoria University of Wellington.

Since 1979 he has been engaged in research into the energy use of existing commercial buildings in the Wellington Central Business District. This work has concentrated on the determinants of commercial buildings' energy use; conservation due to higher energy prices and oil rationing; and detailed monitoring of the energy flows in several buildings to determine energy end-uses. Mr. Pool has co-authored over 15 technical papers and final reports and has given many presentations and lectures on energy related topics.

Bill (W.D.S.) Brander, M.Sc., is a Regional Energy Planner with the Auckland Regional Authority, Auckland, New Zealand. He has responsibility for monitoring and assessing energy developments affecting the Auckland Region and is developing energy policies for the Auckland Regional Planning Scheme.

Mr. Brander received a B.E. degree in Electrical Engineering and an M.Sc. degree in Resource Management from the University of Canterbury, Christchurch, New Zealand. He specialized in energy management and prepared an energy conservation plan for the campus covering administrative, technical, and operational methods of reducing energy consumption. While with the Energy Research Group at the School of Architecture, Victoria University of Wellington, he worked on the development of energy management methods for government buildings and investigated the factors influencing energy consumption in individual buildings. He has co-authored ten papers on these subjects with members of the Group.

Chan Seong Aun, B.BSc., was among the first group of students to receive the new degree of Bachelor of Building Science (B.BSc.) from the School of Architecture at Victoria University of Wellington. He subsequently graduated B.Arch. (Honors) and spent 2 years (1979 to 1981) as a Research Fellow with the Energy Research Group. During that time, he conducted research into energy consumption in schools and office buildings, and into the capital energy requirements of domestic buildings, authoring several papers and reports on the results.

Following on from his research work he spent 2 years with the Public Works Department of Malaysia, as project architect for the Tun Ismail Atomic Research Centre of Malaysia. He is now working with a firm of architects in Kuala Lumpur, as project architect for a multistory office/shopping complex. He is a registered practicing architect in Malaysia.

ACKNOWLEDGMENTS

This book is a work of joint authorship; each author enjoyed the mutual support, comments, and criticisms of his co-authors. Inevitably, joint authorship has increased the number of individuals and organizations who have assisted us in the preparation of the book, and the risk of their omission from this acknowledgment. We sincerely thank all those who did help but who are not specifically mentioned here.

Several organizations provided funding for the research which led to this book. Our main thanks must go to the New Zealand Energy Research and Development Committee, the Ministry of Works and Development, the Ministry of Energy, and the Department of Education. Without their substantial financial support and continued interest in how energy is used in actual buildings, none of this would have been possible. We should also like to acknowledge the significant funding received from the Building Research Association of New Zealand, the University Grants Committee, Victoria University of Wellington and the Standards Association of New Zealand.

We are indebted to the owners and managers of the many buildings we have studied and to the large number of researchers, energy suppliers, building managers, and government officials throughout the world who gave so freely of their time and, just as importantly, their hard-won data on energy consumption in buildings.

No undertaking of this kind escapes the welcome scrutiny of all of our colleagues at the School of Architecture. We must particularly acknowledge Gerd Block, Helen Tippett, and Kit Cuttle for their encouragement and assistance.

We must also acknowledge the special role of Garth Harris, Executive Officer of the New Zealand Energy Research and Development Committee, who first suggested that we attempt this exercise; together with Harry Trethowen, Larry G. Spielvogel, and our colleague Harry Bruhns who gave us detailed constructive comments on the early drafts.

We acknowledge the following individuals for their sterling efforts in helping to produce the final manuscript: Marilyn McHaffie, Linda Searle, Mary Hayward, and Judith Pope for typing; Diana Braithwaite and Gavin Woodward for graphics; Belinda Marshall and Robyn Service for checking and collating.

Finally we must gratefully acknowledge our families and friends for their encouragement, patience, and inspiration.

TABLE OF CONTENTS

Chapter 6
Designer Concerns — Systems Energy Consumption

Chapter 7
User Concerns — Energy Management and Analysis

Chapter 8
Building Energy Performance — Future Concerns

Chapter 1

ENERGY AND BUILDING — EVOLVING CONCERNS

I. INTRODUCTION

This book addresses some of the more important aspects of the energy performance of commercial and institutional buildings. The central concept of the book is that energy performance is the concern, not only of building owners and designers, but also of national governments and building users. This concept provides the framework for the book as a whole, and is described in more detail in Chapter 2.

Implicit in the concept is the assumption that those concerned will wish to improve the energy performance of the buildings for which they are responsible. There is some evidence that this assumption is gaining validity. It is also clear that there are still many impediments to the translation of these wishes into positive action. Many of these impediments relate to existing practices in the financing, owning, designing, construction, and use of buildings.

We have two main aims. The first is to explain the consequences of current practices, as revealed by studies of the energy performance of existing buildings. The second is to provide a basis from which new practices may be developed and new methods of assessing the energy performance of a building may be produced.

Of course, concern with the energy performance of buildings is not a recent phenomenon; it is one which has exercised mankind's ingenuity since time immemorial. This book can only be a snapshot in what is a rapidly changing scene. It can be neither exhaustive nor definitive, but rather provides an in-depth picture of the state-of-the-art in building energy performance (in keeping with the concept of the CRC Press Uniscience Series).

This opening chapter deals with the evolving concerns of energy and buildings. The first section outlines the history of the part played by energy considerations in building design and reviews current thinking on the subject of building performance in general. The second section explores elements of the concept of building energy performance; how to measure it, factors affecting it, and so on. The third and final section reviews theoretical models of energy performance and identifies the relevant interest groups.

II. HISTORICAL PERSPECTIVE

Before attempting to deal with the idea of the energy performance of buildings, it is useful to explore the two main elements of that idea: on the one hand, energy and buildings; on the other, building performance.

Energy considerations have influenced the design and operation of virtually all buildings since time immemorial and it is important to see current studies in context. More recently, the building performance concept has been used in an attempt to provide a basis for the appraisal and specification of buildings. The merging of these two elements into models of building energy performance will be described later.

A. Energy and Buildings: A Brief Historical Introduction

Present day buildings tend to be energy dependent to such a degree that without it they could not be operated or inhabited. Energy is primarily used for the heating, ventilating, cooling, and lighting of buildings. Secondary uses include hot water service heating, vertical transportation, etc. The amount of energy actually consumed depends on the design of the fabric of the building and its systems and how they are operated.

The opportunity for such a high level of energy dependence was not available in the past,

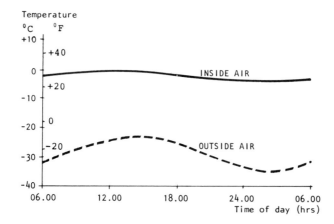

FIGURE 1.   Daily variation in the inside air temperature of an igloo (at sleeping platform level) under typical arctic temperature conditions (From Fitch, J. M. and Branch, D. P., *Sci. Am.*, 203, 6, 1960. Copyright 1960 by Scientific American, Inc. All rights reserved).

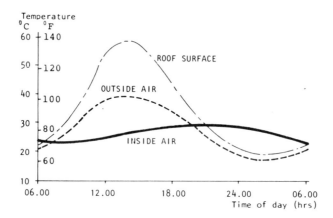

FIGURE 2.   Daily variation in the inside air temperature of an adobe house under high daytime temperature conditions (From Fitch, J. M. and Branch, D. P., *Sci. Am.*, 203, 6, 1960. Copyright 1960 by Scientific American, Inc. All rights reserved).

and primitive man has left abundant evidence of energy-conscious design and construction. Clearly, energy efficiency is a vitally important design criterion when fossil fuels are not available. In many parts of the world primitive man coped (indeed survived) by developing building designs that made use of the available material resources in such a way as to gain the maximum benefit from the climatic conditions.

The igloo and the adobe house are two archetypes of such designs. According to Fitch and Branch[1] ". . . it would be hard to conceive of a better shelter against the arctic winter than the igloo." Figure 1 shows the remarkable temperatures achieved within an igloo, with the occupants and their oil lamps as the only available sources of heat.

Figure 2 shows what is feasible at another end of the climatic scale, with a well-designed adobe house. In this instance, the wide daily external temperature swing is reduced, and the effect of the high daytime peak delayed until late evening. In his analysis, Knowles[2] graphically illustrated how this building type was designed to obtain maximum benefit from winter sunshine, while minimizing summer heat gains (see Figure 3).

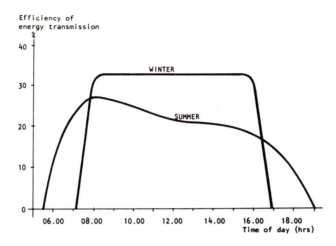

FIGURE 3. Daily variation in solar energy transmission at Pueblo Bonita, N. M. (11th Century A.D.) under summer and winter conditions. Notes: (1) Efficiency is given by the expression (E_t/E_{mt}) per cent where E_t = energy transmitted to the interior of the pueblo and E_{mt} = transmitted energy if all the surfaces of the form were normal to the sun's rays. (2) The winter efficiency is 33% (from 08.00 to 16.00 hr) by comparison with the summer average of 22.6%. (3) The summer efficiency has the desirable characteristic of an early morning peak followed by a gradual reduction as the day progresses. (4) The vertical sun facing walls have high thermal transmission coefficients and high heat storage capacities, by comparison with the horizontal roof structures. (From Knowles, R., *Energy and Form: An Ecological Approach to Urban Growth*, MIT Press, Cambridge, Mass., 1974. With permission.)

Other examples of primitive man's awareness of climatic design principles include the structural forms that were used to cope with tropical climatic conditions of high solar radiation and heavy rainfall combined with moderate temperatures, and the variations that have evolved to cope with subtropical conditions. Fitch and Branch[1] are at pains to point out the "delicacy and precision" of these solutions which must have required "real analytical ability" of the designer.

By comparison, they suggest, Western man consistently underestimates the environmental forces of nature and overestimates his own technological capacities. Figure 4, for example, gives the results of tests on indigenous mud-brick buildings in Egypt and Oman in comparison with modern concrete buildings in the same locations.[3] Figure 4A illustrates the capacity of mud-brick to mitigate the outside temperature. By contrast, the inside air temperature of the concrete building (Figure 4B) exceeds that outside, and the temperature swing is several times that of the mud-brick building.

There appears to be very little published quantitative analysis (of the type presented in Figures 1 to 4) of the performance of the buildings constructed by the Greek, Roman, and other preindustrial civilizations. However, there is ample evidence that designers of these times were well aware of climatic considerations and would apply energy-conscious principles where necessary. This does not imply that earlier generations always gave a high priority to climatic considerations when designing buildings. Then, as perhaps now, "sociocultural factors" were of fundamental importance, with "modifying factors" such as climate, construction, materials, and techniques having a lower order of priority.[4]

The Greeks were well aware of the solar design principles applicable to their latitude and temperature conditions. Individual houses had openings oriented south to allow sun penetration in winter, but were appropriately shaded to keep it out in summer. Urban planning practices were evolved so that individual buildings could utilize solar radiation.[5]

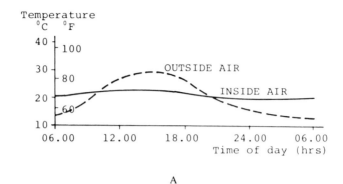

A

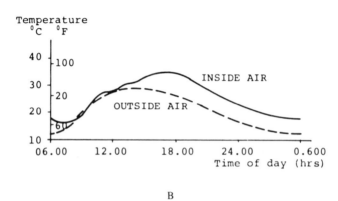

B

FIGURE 4. Comparison of the inside air temperatures in a mud-brick room and a prefab concrete room. (A) Mud-brick room. (B) Prefab concrete room. (From Cain, A., Afshar, F., and Norton, J., *Architectural Design*, 4/75, 207. With permission.)

The Romans were also fairly inventive when it came to the design and application of heating and ventilating systems in their buildings. Underfloor warm air heating systems had been employed to heat bath houses and villas, but the furnaces required large quantities of wood or charcoal. In the face of the resulting shortages of firewood the Romans adopted many of the design principles previously used by the Greeks. In practice, these were adjusted to take account of the wider climatic range embraced by the Roman empire, modified to take account of new developments such as the glazed window, and enhanced by the use of such devices as solar-absorbing floor coverings.[6] The writings of the 1st century B.C. Roman architect Vitruvius embody many recommendations that relate directly to energy-conscious architectural design.[7]

Little further development appears to have taken place until after the Dark Ages. Towards the end of that period the manorial system had taken hold over much of Europe and life, in winter at any rate, centered around the fire in the hall.[8] At that time, a hole in the roof was the only provision for exhausting the smoke. It has been suggested that it was during the Little Ice Age which occurred during the 13th and 14th centuries, that the chimney was increasingly used in buildings. As a result, smaller rooms could be readily heated and the communal life of the great hall with its central fire gave way to a more segregated style, both of life and of architecture. The influence on building design, of the use of the chimney, has lasted to this day.

However, it was not until the approach of the industrial age that energy considerations began to impact on building design in further ways. The availability of cheap and apparently plentiful fuel supplies, together with the development of suitable devices by which the heat of the fuel could be transferred to the interior spaces of buildings, enabled significant changes in architectural practice. The subsequent spread of mechanical ventilation, cooling, and artificial lighting systems thus freed, apparently, the building designer from the constraints of climate. Provided that adequate fuel supplies were available, buildings could be located in any climatic region, as long as sufficiently large capacity boilers and chillers were installed. It should perhaps come as no surprise that this new "freedom" was exploited to the full by building designers, possibly reaching its climax in the glass skyscrapers of the 1960s and early 1970s. Lawrence[9] found that the average energy consumed in New York office buildings constructed between 1965 and 1969 was much higher than that of older buildings.

These energy-intensive buildings contributed to an era of energy growth. According to some authorities[10] we now seem to be at the beginning of a transition from growth to steady state. One of the outcomes of that transition is a desire on the part of many of those professionally involved with buildings to design, construct, and operate them to be as energy efficient as possible.

In a sense we have come full circle and, like our forebears, will be forced to examine carefully the energy consequences of building design and operating factors. With the limited forms of energy and the technologies available to them, our forebears were forced to take a fairly defensive approach to these matters. In many instances, their very survival depended upon the success of their designs, and some of their solutions show considerable ingenuity.

We are now in a position to take a much more positive and scientific approach to the problem. We can give a more appropriate priority to climatic considerations and thus design energy-conscious buildings which minimize demands on nonrenewable resources and at the same time provide a better degree of natural conditioning than was previously possible. Our intention is that the material contained in this book will assist those concerned to take such an approach.

B. Current Approaches to Building Performance

Having briefly looked at some of the historical background to energy and buildings, we now turn to the newer field of building performance.

1. The Lack of a Theoretical Basis

The concept of building performance can be approached from several directions. Buildings are complex systems and it is not unexpected that they can be examined and specified from a variety of viewpoints. It will be evident that such considerations must form a fundamental part of the theoretical basis of architecture. Unfortunately, the theoretical basis of architecture is an area which has received very little attention, at least as far as the building as a whole is concerned. There is no shortage of attempts to develop descriptive and predictive theories,[11] but these tend to be concerned with specific elements of the building. Hence, the prescriptive methods that stem from these theories are also concerned with specific elements of the total building.

Thus it is difficult to place the concept of building performance within the framework of a generally accepted theory of architecture. The concept of building performance is one which has only recently received any systematic study. The term "building performance" implies that buildings can be defined and their performance assessed; but what is a building, how can its performance be measured, and against what is it to be compared? Even the briefest scrutiny of this field immediately reveals the very great difficulties likely to be involved in transforming the concept of building performance into a useful tool for the assessment of existing buildings and the specification and design of new ones. Nevertheless, attempts to devise a single measure profile of building performance are being made.[12]

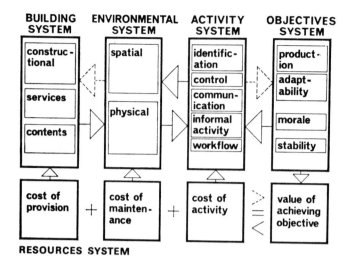

FIGURE 5.   The BPRU conceptual model of the system of building and people. (From Building Performance Research Unit, *Building Perform-ance*, Applied Science Publishers, London, 1972. With permission.)

2. The BPRU Model

The investigations undertaken and reported by the Building Performance Research Unit (BPRU) at Strathclyde University were in the vanguard of systematic studies of building performance.[13] Much of its work concentrated on the development of methods for appraising existing and proposed buildings. One of its more significant achievements was the development of a conceptual model for buildings and the people associated with them. In the context of a multidisciplinary program this model, jocularly termed a "super-mongrel" by BPRU, provided a useful framework for its research.

Figure 5 illustrates the model which the members of the BPRU found useful in the development of their research program. The model is comprised of five main systems, viz:

- **The objectives system** — which, broadly speaking, means the objectives of the organization housed in the building.
- **The activity system** — this system includes all those activities which contribute to the achievement of the organization's objectives.
- **The environmental system** — in other words, the spatial environment (which relates to the dimensional properties of spaces) and the physical environment (heat, light, sound, texture, smell, etc.).
- **The building system** — comprising the construction and services systems necessary to provide the required spatial and physical environment, together with the equipment needed for the activity systems.
- **The resources system** — representing the capital cost of the building system, the maintenance cost of the environmental system, and the operating cost of the activity system which, taken together, would normally be set against the value of achieving the objectives of the organization.

It will be clear that this model provides a general framework for the investigation of other aspects of building performance and indeed for the study of architectural method itself.[14] In terms of building "energy" performance, energy is involved in the capital cost of the building system and in the running costs of the environmental and the activity systems.

In the context of the development of its model, it is interesting to note the concern of the

BPRU, not only with the building, but also with all the people involved, the "... designers, owners, users, and producers of buildings." To some extent these early ideas presage the conceptual basis of this book, which takes the concerns of all the various groups of people likely to be involved with buildings as its central theme.

3. The Four Function Model

Hillier and Leaman[15] explored the need for better theoretical and methodological concepts in the broad field of architectural research. While it seemed to have become feasible to achieve some of the individual requirements (e.g., thermal, visual, acoustic) of building performance, it was evidently not possible to satisfy a large number of these simultaneously. The individual requirements had been fairly thoroughly investigated, but there had been little research on their integration in design, a fact to which many failures in building performance were attributed.

Their conceptual model was intended to meet the specific needs of architectural research. However the model has implications for building performance. The key idea was that the "building problem" should be divided into the different functions of the building, rather than the divisions that result from the contributions of the traditional professional disciplines. Four general functions were identified:

- **Climate modification** — which ranges from physiological needs through technical to microclimate conditions, and their integration.
- **Behavior modification** — concerned with the relationship between human activity and its spatial containment.
- **Resource modification** — buildings as users of resources.
- **Cultural modification** — related to the symbolic function of buildings.

These functions were conceived as representing ways of thinking about the building as a whole, rather than parts of it as is current practice. While the concept of the energy performance of buildings has implications for all four functions, it will be evident that it relates most strongly to the resource modification function.

4. The Performance Approach

The use of the so-called "performance approach" has been advocated in connection with many aspects of building: preparation of design data and guidance, project design and construction, product development, quality control, and so on. The performance approach[16] is defined as the practice of thinking and working in terms of ends rather than means and is concerned with what the building (or building product) is required to do, rather than prescribing how it is to be constructed. Ideally, performance requirements should be defined in quantitative terms, and methods should be available to objectively evaluate each aspect of performance.

Data deficiencies prevent the full exploitation of this approach at the present time. However, as will be described in Chapter 2, a start has been made towards the quantitative specification of building energy performance requirements. Chapter 6 will deal with evaluation at the prediction stage, Chapter 7 with performance in use.

The performance approach is receiving increased attention in connection with building regulations and standards. Traditionally, many standards have been prescriptive in nature. However, with the ever widening range of means of construction available to the designer and the increasing complexity of attempting to simultaneously meet several (sometimes conflicting) prescriptive standards, national and international standards bodies have been examining the possibility of using the performance approach. The main thrust of this activity has been towards predictive or design performance standards, but standards which relate directly to the actual performance in use are also being looked at. The question of energy performance standards for buildings will be discussed in detail in Chapter 3.

III. ASSESSING BUILDING ENERGY PERFORMANCE

In this section we shall briefly outline some of the background to the assessment of building energy performance, consider why such assessment may be of value, and explore how it might be carried out.

Although not a new factor in the design and operation of buildings, there is no doubt that energy principles are going through a process of rediscovery at the present time. Energy performance criteria were implicit in the buildings of primitive man, who had little choice in the matter if survival was to be possible in other than a few benign regions of the world. However, during the recent period of energy growth, postindustrial man appears to have lost sight of such criteria. The result has been an apparent fall-off in the energy performance of buildings, the reasons for which were perceived by Stein well before the current energy crisis and elucidated in *Architecture and Energy*.[17]

According to Dumas,[18] in his review of energy-using systems of all kinds, ''. . . the single most important determinate (sic) of the life-cycle energy performance of any building is its architectural design.'' Almost in corollary to that statement, Hawkes[19] has expressed the conviction that ''. . . energy conservation is an important part of the current debate about the nature of architecture,'' while Caudill is reported[20] as seeing ''. . . energy as a fifth category of principal design considerations, along with function, form, economy and time.''

As mentioned earlier, explorations of the general concept of building performance are still in their infancy, while studies of building energy performance are even more recent and poorly defined. To be sure, the bases for assessing the energy performance of many building components and individual systems have been fairly well established. Here, however, our concern is with the energy performance of the building as a whole, which involves the complex interaction of the components and systems of the building and consideration of their combined dynamic behavior.

Of course, the definition or quantification of energy use does not in itself give an indication of performance; the use of energy must be related, for example, to some measure of the building size or the activities it houses or the objectives of the organization using the building. In terms of the BPRU model described earlier (see Figure 5) one could assess energy performance both in relation to the provision of the building system and in respect of the maintenance of the environmental system. Potentially, it is feasible to assess the energy cost of the activity and objectives systems too and thus quantify the energy performance of the organization as a whole,[21] but that is beyond the scope of the present work.

 Assessment of energy performance is important at the design stages, during construction, and throughout the life of a building. Such assessments may be used as a measure of the relative energy performance of the building stock of nations or simply as a yardstick for comparing two buildings containing similar activity systems. Energy performance criteria may also be used to check design predictions or as a basis for relevant standards. Specific applications will be explored further in Chapter 2 and discussed in detail in the chapters which deal with the relevant concerns of nations, building owners, designers, and users.

In terms of measurement, the energy use of a building would appear to be one of the more easily quantifiable measures of building performance. Unfortunately, many practical difficulties arise when one attempts to develop a consistent measure of energy performance. These difficulties range from choice of units to concerns with the use of indigenous and of nonrenewable energy resources. Questions of the weighting to be given to different types of delivered energy, with respect to the source of fuel from which they were derived, are yet another headache for those who would attempt to measure energy performance with precision.

In practice, these difficulties are not insurmountable. What tends to be much more difficult is achieving a consensus on what standardizing factors to use in order to provide a consistent measurement of energy performance. Conventional practice (insofar as so recent a field has developed conventions) is to use a determinant which is in some way related to the facilities provided by the building. This could be a measure of floor area of the building or some other size-related dimension such as building volume. Alternatively, some determinant more closely associated with the provision of the building for a specific activity may be used. For example, energy use per hospital bed, per school pupil, per office worker, per hotel room, and so on are all possibilities in the appropriate circumstances. Sometimes, energy use may be linked directly with the activity system; for example, energy use per unit of sales in the case of retail stores, energy use per meal in the case of restaurants, etc., in a manner not dissimilar to an industrial production process.

The energy performance of a building is affected by many factors. The influence of climate has already been mentioned and its effects, together with those of activity and building size, will be discussed in detail in Chapter 4 under the heading of "Owner Concerns." The important influence of design factors has also been stated; these will be dealt with in Chapters 5 and 6. The factors affecting the energy performance of the building when in use will be outlined in Chapter 7, in the context of the management and analysis of energy use.

The next step is to examine the available models that relate to the energy performance of buildings, as background to the thinking behind the conceptual framework on which we have structured this book.

IV. MODELS OF BUILDING ENERGY PERFORMANCE

Energy use is tied up in many aspects of building design, construction and operation, and involves the input of several professional disciplines. It is no surprise, therefore, to learn that a range of energy performance models has been developed relating to the requirements of these various aspects of building and from the viewpoints of the different disciplines. For the purpose of this section, these have been divided into engineering and architectural models. More recently, generalized energy models have been proposed which attempt to cross the boundaries embodied in the above classification and take into account the other factors required for a complete modeling of building energy performance.

These three types of model — engineering, architectural, and energy — will now be reviewed. The reviews will be purely descriptive. Their development for use in the analysis, design, prediction, and management of building performance will be detailed in later chapters.

A. Engineering Models

Engineering models may take various forms but all of them are concerned with the energy-consuming services of buildings. At their simplest, they deal with a single building and consist of a simple listing of the energy-consuming functions. Dubin and Long[22] itemize the following such functions: heating, lighting, cooling, power for equipment, and domestic hot water. These functions may be diagramed as indicated in Figure 6 which is a schematic of the principal energy systems and flows in a "conventional" air-conditioned building.[23] A fossil-fueled central boiler and an electrically driven chiller supply heating and cooling energy to a central air-handling plant which distributes air within the building. Direct energy losses occur via the boiler flue, the cooling tower, and the exhaust air. In the example shown, electricity is also used by motors (for fans, pumps, dampers, elevators, etc.), lights and miscellaneous equipment, and for domestic hot water heating.

All the energy converters within the building (even though their primary function may be

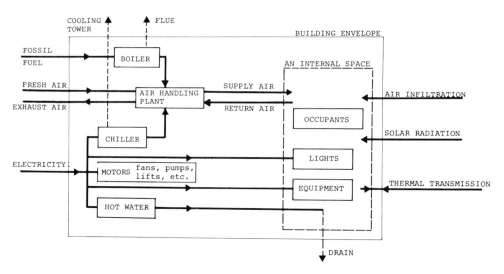

FIGURE 6.   Schematic of the principal energy systems and flows in a conventional air-conditioned building.[23]

to provide light or mechanical energy) ultimately produce thermal energy which raises the temperature of the air and of the interior surfaces of the building. This process applies equally to energy-consuming equipment and to building occupants; ten people produce more heat than a 1-kW electric heater.

Energy transfers, both within the building and between it and the external environment, occur by distinct processes, some of which are indicated in Figure 6. A very important energy transfer occurs when the radiant energy of the sun passes through a window and is absorbed on interior surfaces where it is converted into thermal energy. If the interior air temperature is lower than desired, this transfer is a free source of energy which helps raise the indoor temperature. However, if the indoor temperature is too high, this transfer is unwanted and can result in the use of additional energy for cooling.

Buildings seldom have all of the energy conversion processes shown in Figure 6. For example, many do not have chillers, some do not have boilers but use local heaters to warm the air directly, and others have opening windows rather than air-handling units. However, most buildings have some of these items of equipment, and it is important that they be included in the modeling process.

The sizing of all these energy-consuming systems is bound up in traditional engineering practices. However, the need to predict energy consumption with greater accuracy has prompted a reassessment of these practices and a drive towards more sophisticated models, not only at the individual systems level, but also at the level of integration of these systems.

As will have been evident from Figure 6, however, integration of the engineering systems is not sufficient. They, in turn, must be integrated with the building system. O'Callaghan[24] makes this point in connection with the analysis of the likely effects of energy conservation measures. Figure 7 illustrates his energy conservation design model for a building, which in his terms is ''. . . a complicated and intricate thermal assembly.''

Maver,[25] as might reasonably be anticipated of a member of the BPRU, locates the design of the energy consuming services within the building system of the BPRU model of building performance. Although much of his investigation relates to the design of the individual subsystems, he also takes up the point of their integration at (and beyond) the single building level. He is particularly concerned with the basic energy forms and their interchangeability, together with the alternative ways of meeting building energy demands. Figure 8 illustrates and summarizes many of these alternatives. Maver explores several of these and discusses their implications. He stresses the need for the interaction between the different services

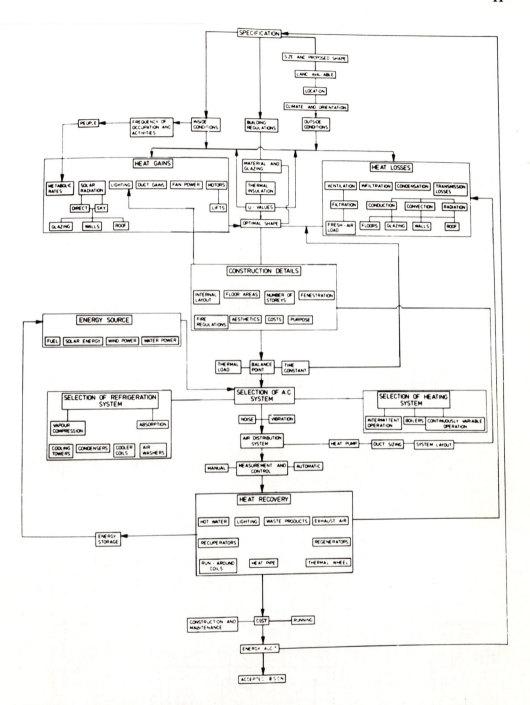

FIGURE 7. O'Callaghan's model of building design for energy conservation. (From O'Callaghan, P. W., *Building for Energy Conservation*, Pergamon Press, Oxford, 1978. With permission.)

and between the services and the building system, to be studied at the earliest possible stage in design. He also stresses that ". . . the study of the energy requirements of the built environment as a whole is properly within the responsibility of the architect/planner."

We shall now examine some architectural models relating to the energy performance of buildings.

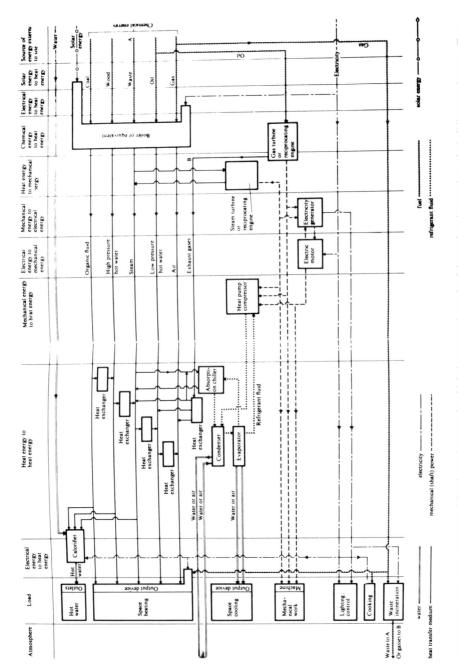

FIGURE 8. Maver's summary of the alternative ways in which the energy demands of the engineering services of a building may be met. (From Maver, T. W., *Building Services Design*, RIBA Publications, London, 1971. With permission.)

B. Architectural Models

Generally speaking, architectural models have been concerned more with the environmental performance of buildings than with the energy consequences of achieving that performance. Many architectural models are derived from related engineering models, but whereas engineering models tend to be involved with the estimation of, say, heating and cooling design loads, architectural models are more usually concerned with the prediction of the inside temperature resulting from a given configuration of the building fabric. One implication of this is that the amount of computation required may often be considerably more, as different architectural options are explored (as will be noted later, energy models may require still more computational efforts for their resolution).

Banham[26] has identified what he termed three modes of environmental management: the conservative, the selective, and the regenerative. The conservative mode involves the employment of relatively massive construction which not only provides thermal insulation but also thermal storage capacity, the net effect of which is to reduce the variation in inside temperature compared to that outside. The selective mode refers to the use of the building fabric too, but here the concern is with features which will admit any desirable climatic elements, while rejecting others: for example, the glazed window which prevents ingress of rain and cold air, but admits daylight and solar heat gain. The regenerative mode implies the use of energy derived from nonrenewable fuels, in devices ranging from open fireplaces to artificial lighting installations and air-conditioning systems.

All three modes are present in most buildings. With only limited regenerative aids at his disposal, preindustrial man concentrated on the conservative and selective modes, the choice depending on the prevailing climatic and cultural conditions. However, over the last century or so, the availability of a range of regenerative aids has allowed the environmental management of building to become much more dependent on energy inputs. Banham's model provides an excellent basis from which to explore this topic further.

In his review of the history of models of the environment in buildings, Hawkes[27] is mainly concerned with thermal, visual, and aural environmental modeling (heat, light, and sound). He traces the relatively long history of the development and use of such models (in design at least). At the time of his writing (1970) he did not explore the energy implications of those aspects of environmental control but, as will become apparent in the next section, both capital and running energy costs are involved in all three aspects; running energy costs usually dominating in the case of heat and light, capital energy costs in the case of sound.

Hawkes also points to the potential application of computers for facilitating a more comprehensive view of these matters, compared to the piecemeal approach necessitated by the manual methods used at that time. The availability of increased computing power, together with the spur of the energy crisis of the 1970s, led in fact to much more effort being put into the design of more comprehensive models.

Still at the descriptive level, Markus and Morris[28] put forward a model of "climate, building, shelter and people" based on the concept of the continuity of space and the second law of thermodynamics. Developed to the level of a serviced building, as indicated in Figure 9, it comprises two overlapping environments. Environment 1 consists of a resource environment and a shelter system, while Environment 2 contains the human system and a resulting controlled environment. The overlap between these represents the controlled environment provided by the shelter system, which is also the resource environment for the human system. The control mechanisms are indicated. Such a generalized model illustrates the links between the building occupants and the use of fossil fuels, within the context of the man-made environment.

We shall now move on to consider what energy models (as defined earlier) are available.

C. Energy Models

Our concern here is with descriptive models of the energy performance of buildings. The

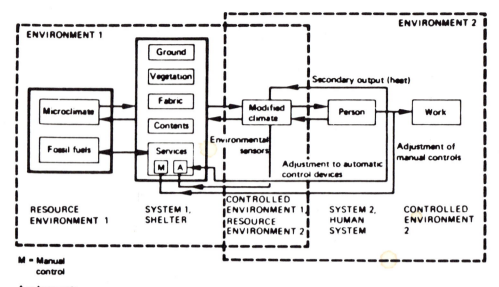

FIGURE 9.   The Markus and Morris model for a point in a serviced building. (From Markus, T. A. and Morris, E. N., *Buildings, Climate and Energy,* Pitman Books, Ltd., London, 1980. With permission.)

detailed development of these models for specific purposes (design, management, etc.) will be dealt with in later chapters.

Stein[17] has used the energy consumption profile of a building over its entire life as a graphic energy model. Figure 10 is based on Stein's concept. It shows the energy consumption profile of a hypothetical building, starting with the energy used during the construction period and ending with that expended in the demolition process. During the assumed 40-year life of the building illustrated, the profile exhibits an initial phase of erratic energy use, followed by a 10-year period with a regular annual consumption cycle, the possible effects of a major retrofit at year 25 and, towards the end of its life, a gradual rise in consumption.

In practice, the shape of such a profile would be influenced by a whole host of factors, some of which are still very difficult to quantify due to a lack of reliable data. The problems inherent in estimating the energy expended in producing building materials and directly in the construction process, that is, "the capital energy requirements" of the building, will be explained in Chapter 5. The factors affecting the energy used directly to operate the building, that is, the "systems energy consumption," will be described in Chapter 6. Suffice it to say at this stage that each building is likely to have its own unique profile.

The use of the type of energy analysis described above is still relatively novel in the field of building energy performance. However, similar concepts underlie the application of conventional life-cycle economic techniques and these have been developed[29] for use in relation to energy conservation. Economic models of this type are used, for example, to determine how much capital should be spent on energy conservation measures, in order to minimize the total cost of the building over its useful life.

As described earlier, engineering and architectural models are available which allow the effects of external climate, the building fabric, and the energy-consuming systems to be taken into account. What is largely missing from these models is the effect of how the building and its systems are used and controlled by the occupants and operators. Spielvogel[30] for instance, has strongly asserted that these human factor considerations are by far the most important in determining the energy performance of buildings. However, they are still largely ignored. One of the purposes of this book is to emphasize, and where possible to quantify, the influence of these human factors.

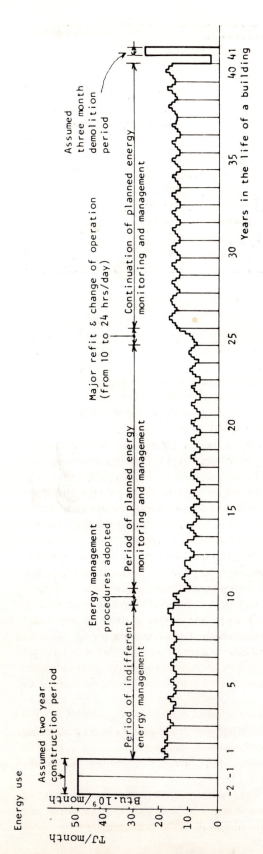

FIGURE 10.   Energy consumption profile for a hypothetical building over an assumed 40- year life span.   (After Stein[17]).

Rubin[31] too, has concluded that the actions of people are a critical determinant of the energy performance of buildings, in contrast to the conventional emphasis on the purely physical aspects of modeling. He also recognized the dearth of empirical data that would allow the quantification of the influence of people; we deal with this matter in more detail in Chapter 6.

In his attempt to model energy consumption in buildings, Brander[32] utilized the three basic systems depicted in Figure 11 and designated as follows: energy systems, nonenergy systems, and human systems. Energy systems are defined as those which use energy directly in providing some useful service such as heating, ventilating, or lighting. Such systems usually involve a piece of equipment, together with associated controls regulating its mode of operation, and are normally covered by engineering models.

By contrast, nonenergy systems do not use energy directly but affect the amount which energy systems consume. These nonenergy systems comprise the building envelope characteristics (e.g., thermal resistance and airtightness), interior surfaces (position and surface reflectances), doors, windows, and furnishings. They affect the flow of ''purchased'' energy from a space or its level of utilization within a space. They also affect the flow of free energy (e.g., solar radiation and outside air) into a building, which may either increase or decrease the requirement for purchased energy. These nonenergy factors will be recognized as those usually included in the architectural models discussed in the previous section.

Both energy and nonenergy systems are controlled by human systems (such as working times, occupant behavior, cleaning schedules, maintenance, and thermostat settings). Administrative decisions determine the various activities which occur in different parts of a building and their schedules. The behavior of occupants determines the effect that use of nonenergy systems (e.g., drawing curtains and opening windows) has on energy systems. It also influences the operation of energy systems where occupant intervention is possible. Maintenance procedures determine the degree to which plant and controls operate efficiently and as intended, and how nonenergy systems will affect energy use. The building occupants themselves constitute a source of heat within a building; they are an energy system and must be taken into account from that point of view.

The ways in which these three systems behave and interact determine the energy performance of a building. All three must be taken into account in any useful energy model.

The importance of each of these systems varies from building to building. In large buildings, with centralized air-conditioning systems and sealed facades, the occupants may have little direct effect on building energy consumption. However, the human component can still have a profound influence on energy consumption. If the building services are largely automated then the personnel responsible for maintaining equipment and setting controls have a key role to play in relation to energy use and energy management. If the building services are under manual control, the building manager assumes the key role.

In buildings with opening windows, personal heating systems, and locally controlled lighting, the behavior of the occupants and their working hours can become the major influence on energy consumption. The users have a considerable effect on energy consumption, and energy management strategies need to be more broad based than in the sealed building described above.

The model of building energy performance employed in our investigation of this field uses concepts embodied in the work of both Stein and Brander. Building energy performance is perceived as a concern at the following levels of the human system: national institutions, building owners, building designers, and building users, and is explored from the point of view of each.

It will be seen that each level of concern provides the environment for the next. The resulting framework forms the basis for this book. We have termed it the exogenous-endogenous framework and it will be explained in detail in the next chapter.

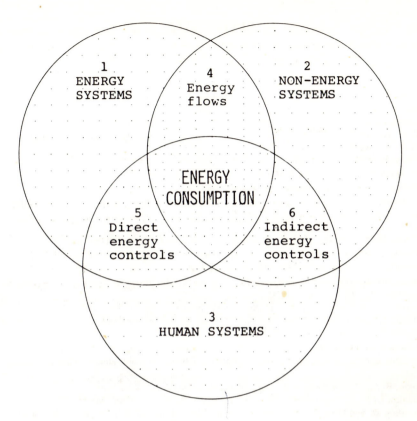

FIGURE 11.   Brander's model of the interaction of energy, nonenergy, and human systems.[32]

Notes:

1. Energy systems	2. Nonenergy systems	3. Human systems
Space heating	The building fabric—	Working times
Mechanical	doors, windows,	Space
ventilation	walls, floors,	utilization
Lighting	roofs, partitions,	Occupant
Equipment	curtains	behavior
Occupants (metabolic	behavior	Maintenance
energy)		schedules

4. Energy flows	5. Direct energy controls	6. Indirect energy controls
Air infiltration	Thermostats	Opening windows
Solar gains	Valves and switches	Drawing curtains
Thermal conduction	Time clocks	

REFERENCES

1. **Fitch, J. M. and Branch, D. P.,** Primitive architecture and building, *Sci. Am.,* 203, 6, 1960.
2. **Knowles, R.,** *Energy and Form: An Ecological Approach to Urban Growth,* MIT Press, Cambridge, Mass., 1974.
3. **Cain, A., Afshar, F., and Norton, J.,** Indigenous building and the third world, *Archit. Design,* 4/75, 207.
4. **Rapoport, A.,** *House Form and Culture,* Prentice-Hall, Englewood Cliffs, N.J., 1969.
5. **Butti, K. and Perlin, J.,** *A Golden Thread,* Cheshire Books, Palo Alto, Calif., 1980, chap. 1.
6. **Butti, K. and Perlin, J.,** *A Golden Thread,* Cheshire Books, Palo Alto, Calif., 1980, chap. 2.
7. **Vitruvius,** *The Ten Books on Architecture,* transl. by Morgan, M. H., Dover Publications, New York, 1960, Book VI, chap. 1 and 4.
8. **Burke, J.,** *Connections,* Macmillan, London, 1978.
9. **Lawrence, C. W.,** Energy use patterns in large commercial buildings, in *Energy Conservation: Implications for Building Design and Operation,* Conf. Proc., Abrahamson, D. E. and Emmings, S., Eds., University of Minnesota, Minneapolis, May 1973.
10. **Odum, H. T. and Odum, E. C.,** *Energy Basis for Man and Nature,* McGraw-Hill, New York, 1976.
11. **Woodcock, E.,** *Descriptive, Predictive and Prescriptive Theories in Architecture,* Course Notes, Scott Sutherland School of Architecture, Aberdeen, Scotland, 1969.
12. **Alexander, K.,** Performance — A Direction for Research, CIB W60, The Performance Concept in Building, Document No. 13/11, March 1980.
13. Building Performance Research Unit, *Building Performance,* Applied Science Publishers, London, 1972.
14. **Daish, J.,** *Architectural Method Handbook,* 2nd ed., School of Architecture, Victoria University, Wellington, 1979.
15. **Hillier, B. and Leaman, A.,** A New Approach to Architectural Research, *RIBA J.,* 517, December 1972.
16. Working with the Performance Approach in Building, CIB Report, Publication 64, CIB, Rotterdam, January 1982.
17. **Stein, R. G.,** *Architecture and Energy,* Anchor Press/Doubleday, New York, 1977.
18. **Dumas, L. J.,** *The Conservative Response,* Lexington Books, Lexington, Mass., 1976.
19. **Hawkes, D. and Owers, J.,** *The Architecture of Energy,* Construction Press, Essex, England, 1981.
20. Energy Analysis Techniques for Building Design, Supplement 80-4 to the AIA Energy Notebook, American Institute of Architects, Washington, D.C., December 1980.
21. **Odum, H. T.,** *Environment, Power and Society,* Wiley-Interscience, New York, 1971.
22. **Dubin, F. S. and Long, C. G.,** *Energy Conservation Standards,* McGraw-Hill, New York, 1978.
23. **Baird, G. and Brander, W. D. S.,** *Energy Conservation Checklists for Occupants and Operators of Government Buildings,* CRP 20, School of Architecture, Victoria University, Wellington, 1982.
24. **O'Callaghan, P. W.,** *Building for Energy Conservation,* Pergamon Press, Oxford, 1978.
25. **Maver, T. W.,** *Building Services Design,* RIBA Publications, London, 1971.
26. **Banham, R.,** *The Architecture of the Well-Tempered Environment,* The Architectural Press, London, 1969.
27. **Hawkes, D.,** *A History of Models of the Environment in Buildings,* Working Paper 34, Centre for Land Use and Built Form Studies, University of Cambridge, September 1970.
28. **Markus, T. A. and Morris, E. N.,** *Buildings, Climate and Energy,* Pitman, London, 1980.
29. **Marshall, H. E. and Ruegg, R. T.,** Energy Conservation in Buildings: An Economics Guidebook for Investment Decisions, NBS Handbook 132, U.S. Department of Commerce, National Bureau of Standards, Washington, D.C., May 1980.
30. **Spielvogel, L. G.,** How and why buildings use energy, in *Energy Conservation through Building Design,* Watson, D., Ed., McGraw-Hill, New York, 1979.
31. **Rubin, A. I.,** Energy Conservation in Buildings — A Human Factors/Systems Viewpoint, NBS Building Science Series 88, U.S. Department of Commerce, National Bureau of Standards, Washington, D.C., November 1976.
32. **Brander, W. D. S.,** Energy Conservation Strategies for the University of Canterbury, M.Sc. Research Project Report, University of Canterbury, Christchurch, 1980.

Chapter 2

BUILDING ENERGY PERFORMANCE — A NEW FRAMEWORK

I. INTRODUCTION

It is our belief that the range of influences on building energy use is so great that a classification system is essential if the different people involved are to properly understand their role in connection with building energy performance. There have been many attempts at such classifications over the years.[1-3] All have addressed the labeling of building and/or environment features to some degree. We believe the framework on which this book is built deals with the problem from a completely different angle: that of the people involved in controlling, buying, building, owning, or occupying a building. Each of these groups of people has a different interest, a different concern, with respect to building energy use.

It must be emphasized that far from being yet another example of academic obfuscation this type of classification has considerable utility. It concentrates attention on the people for whom the building was built and for whom energy is used inside the building. Most importantly the classification highlights the areas where concerned people can actually have some influence. For this last reason, the chapter labels refer to the people who may be "concerned" and in this context their "concerns" are purely in those areas where they may make effective decisions.

This chapter introduces the overall energy performance classification system. It is based on the concept (borrowed from ecology) of exogenous and endogenous influences on systems. The place of energy performance indexes within this scheme is then discussed in detail.

After describing the energy performance classification system itself, we present a two-part discussion on indexes of energy performance. The first part examines the generation of indexes, the variables used to standardize energy use, and the complications that can arise in the formation of energy use indexes. The second part deals with the applications of performance indexes and the forms of index that suit these applications. Such indexes are essential for the establishment of useful measures of building energy performance.

Taken together these two parts set out the theoretical background to the various measures of building energy performance which will find application in the overall framework of the book. In effect, this exogenous-endogenous (exo-endo) framework provides the units for the measurements, while the indexes provide the numerical value of those units; within a particular classification of energy-use-influencing parameter, one or other of the indexes provides the best measure of the actual size of that influence.

II. THE EXOGENOUS-ENDOGENOUS CLASSIFICATION SYSTEM

A. Definition

The two words "exogenous" and "endogenous" have the following dictionary definitions: exogenous (eksojenus) adj., originating outside an organism; endogenous (endojinus) adj., (biol) growing or originating from within an organism. When contrasted in this way, these definitions express very clearly the critical concept of the classification scheme for building energy performance that forms the basis of this book. The chapter structure is based upon the exo-endo scheme; indeed, the terminology of our classification of readers of this book is structured to fit the exo-endo concept. At base the concept provides a framework for the classification of building description factors (parameters) according to their position in a hierarchy. The hierarchy is structured according to the type of people who can influence the values of these factors.

In the simplest case, the energy consumer (the "user" of the energy) normally has less influence on the design or construction of the energy-consuming services than the designer has. The user is subjected to a range of exogenous factors like heating system type, transportation system, or lighting arrangement, which are in fact endogenous factors for the designer. The user's endogenous factors (that might affect their use of energy) are related to the management of these, given, energy-consuming systems.

The classification looks at constraints on and options open to the various parties involved in buying, building, owning, or occupying a building. The constraints are the exogenous factors for a particular party, while the options available are the endogenous factors about which that same party may make decisions.

It is a principal strength of this classification system that, in concentrating on the action that people involved with a building may take, it maximizes emphasis on the purpose of energy use in a building. It is unfortunate that phrases such as "building energy use" or "energy consumption by buildings" have become common because they have tended to distract attention from the real purpose of a "building's energy use", that is, the provision of services for the occupants or users of that building. The exo-endo classification concentrates attention on the users of energy-consuming systems and classifies the various features of a system according to its potential for modification. At a given level, exogenous features are external constraints on the energy-consuming systems, while endogenous features are aspects that are amenable to change.

The focus on people and their potential for action does not stop at this point. Beyond questions about energy use "in" or "by" a building, the exo-endo classification system identifies particular levels of people's interest in energy system features. It recognizes that what is a constraint (i.e., exogenous) for some people may in fact be a choice (i.e., endogenous) for others.

In putting forward a further intellectual framework for building energy use studies it is essential that the reasons for selecting it, ahead of those already available, be put forward. The preceding paragraphs set out the preliminary arguments in such a case, the principal advantage of this framework being that it emphasizes end-uses and identifies action levels, a major shortcoming of other classification systems. In brief, the principal strengths of the exo-endo classification system are that it identifies not only the factors that affect energy consumption, but also the people or organizations who exert some control on those factors, and hence influence that consumption.

B. Existing Classification Systems

It is possible to find alternative models of building energy use. The architectural models of Chapter 1 are such alternatives. As with the exo-endo classification system many of these models classify the factors affecting energy use in a building. The labels of the classification serve to clarify the general pattern of interrelationships at the heart of the operation of a building.

One such model sees the building acting as an active filter between the external and internal environments. Dependent on the "settings" of parts of the filter (like the solar blind angle) the impact of variations in outdoor climate, on indoor comfort or performance, may be large or small. For example, large variations in outdoor temperature can be mitigated by the filter so that the corresponding temperature variations indoors are small and as close as possible to human comfort conditions.

This "building as a filter" approach concentrates attention on the building fabric. Energy-consuming systems merely provide the fine tuning to the performance already provided by the building fabric. We would subscribe to the implied philosophy of this approach, as it sees sensible building design as the crucial first phase in the production of low energy buildings. We would suggest, however, that this approach overemphasizes the reduction of energy loads, the conservation side of the equation, and ignores the other side, that of energy supply. It is not sufficient just to minimize the need for energy-consuming services, it is also necessary to consider the efficient and effective use of the energy that is used, in the fine tuning. Both these essential questions are examined in more detail in Chapter 6.

The focus on the "building as a filter" as the key element in this model also detracts from the emphasis that we feel must be placed on people, in the examination of building energy performance. Concentrating on the properties of the building fabric diverts attention away from the needs and behavior of the people in the building. For example, designers may concentrate on the performance of the windows rather than the window-people systems of their buildings. Their optimum design can easily be subverted by unanticipated occupant behavior. Chapter 7 examines the building energy management aspects of this behavior in detail.

The exo-endo classification avoids these problems by concentrating on the people involved in creating, altering, and using a building and, more especially, on their roles as determiners of energy use. In the same way that classical Newtonian physics forms a valid subset (with restricted application) of relativity theory, the "building as a filter" model forms a valid subset of the exo-endo classification used in this book.

C. The Exogenous-Endogenous Classification in Detail

Figure 1 illustrates the exo-endo classification and its various levels. The first point to emphasize is that it allows the systematic study of energy use in buildings. It is a classification system with which to order, or organize, information about energy use in buildings. Primarily, it will be used to classify various factors about a building (or set of buildings) that influence their energy use. The exo-endo classification has been used in other contexts and for other applications. Its use here should not be confused with these other applications.

The basic idea for the classification of systems with these exo-endo labels comes from ecology. Ecological systems are classified by the degree of control exerted on them from within the systems themselves. Endogenous systems have internal control, while exogenous systems are controlled from outside.

These labels have already been applied to buildings, in a manner much more closely paralleling their origins in biology. Markus drew an analogy between building structures and biological structures.[1] The traditional building with large masonry load-bearing walls could be seen to have an "exo-skeleton" in the way that lobsters do; the structure of the organism comes from outside, from its shell. Frame plus curtain-wall structures were seen to possess an "endo-skeleton" much like horses or people; the structure of the organism this time comes from inside, from its skeleton. The primary advantage for this view of a building was seen to be contained in the notion of the building as an organized system or organism; " . . . that is, having a complete set of inter-related parts and subsystems, controls, feedback devices, means for energy storage and generation, and the dynamic response to change such that desirable homeostasis (as in the body) can be attained."[5]

Our intention is to extend the "ecological" definition into the energy field. This will allow the study of buildings as a part of a social system for which, at a given level, there are various factors which are exogenous (external) and hence not able to be controlled, and various other factors which are endogenous (internal) and hence able to be controlled.

1. The Nation

In the first instance, our model takes the nation as the system. We could have gone further and looked at "society", "the world", "spaceship earth", and so on up to the rarefied levels more usually occupied by theologians and philosophers. We resisted that temptation. It was apparent that the most effective action relating to buildings would be taken at the national (or lower) levels.

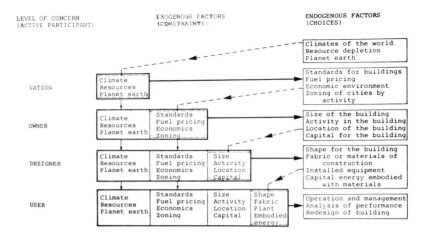

FIGURE 1.   The exogenous-endogenous classification system for factors which affect the energy performance of buildings.

The model establishes a hierarchy ranging from national concerns through building owner and designer to user concerns. At the national level there are few (exogenous) constraints and (given sufficient regulations and incentives) many (endogenous) factors that may be controlled. At the user level the opposite is true: much is exogenous, and little can be controlled or varied (is endogenous).

Even at the national level however there are a number of exogenous factors that affect energy use in buildings. Aspects of energy consumption such as energy resource availability (e.g., type of coal, prices of imported fuel) climate, and natural energy availability (e.g., useful solar and wind energy) are fixed exogenous factors for a nation. There will be variations in climate from one location to another, but these variations are fixed for a particular nation. Similarly, the resource picture for a nation is exogenous, despite the variations brought about by mineral, oil, and other such explorations and discoveries.

Endogenous factors for a nation are those aspects of a building about which a particular society or government may make regulations or offer incentives. A nation's concern is often expressed by compulsory standards for the performance of a particular component or building. It is possible to imagine at this level that all the (for want of a better phrase) lower level endogenous factors could be brought up to this level: the nation could decide how people must use buildings. The difficulty that remains is in conceiving of an enforcement system for this mandated behavior. The decision still remains in the hands of the user (at the "user" level) whether to obey the government strictures or to avoid them. The management of the energy-consuming systems is still an endogenous factor for the user, despite government decisions, and can be used to avoid the spirit of a regulation even while technically complying with it.

The national level is where the economic structures are set, within which the various other "levels" of "active participant" operate. The energy decisions of the owner, the designer and the user, the three other levels of the classification system, are constrained by the decisions of the nation on economic matters. For example, the fuel or resource prices controlled by a government become, as do all decisions about national concerns, constraints on decisions at the other levels. Financial incentives are likely to be one of the more effective expressions of national concerns. Such incentives (the "economic environment" and "fuel pricing" of Figure 1) become the constraints for all other levels of concern.

2. The Owner

As expressed by the reappearance of endogenous factors (choices) as exogenous factors (constraints) at each "lower" level in Figure 1, the choices of the nation become the constraints at the lower levels of concern. Specifically, in moving from the second to the third level in Figure 1, the exogenous factors of the nation, plus any decisions (endogenous factors) made by the nation, become constraints for the building owner. In the exo-endo model, the nation is the "creator" of many of the exogenous factors at the owner level. Continuing on to the next level, the owner becomes the creator of the exogenous factors at the level of a designer's interest in a building; finally, for the user, the exogenous factors become those decisions made by designers plus all the designer's exogenous factors.

Therefore, in the construction or purchase of a building, when the owner is concerned and indeed has some control over decision making, the exogenous factors are the resource pricing, economic planning, activity zoning, and standard-making decisions that have been made at the national level, plus the constraints of national climate and resources availability. The endogenous factors for the owner can be split into three major areas: the location of the building, the activity it is to house, and the scale or size of the building to be purchased or constructed. Depending on the owners' knowledge and ability to act, they will exert a greater or lesser degree of control on the design of the building. The greater the control, the larger the number of endogenous factors at the owner level. For the purpose of this book, however, we have divided what is normally considered of concern to the owner of a building from what is usually considered of concern to its designer.

The location of a building, its size, and the activity it houses, are of primary concern to the owner. It is the owner who has decision-making power at this level. Whatever decision is made beyond this by the owner, about window size and placement or about heating system type, for example, is likely to be influenced by the designer. Such decisions are predominantly the concern of the designer rather than the owner.

For the owner of a building, whether financier, developer, constructor, investor, or lessor, the location of a building can be an extremely important consideration with respect to energy use. In addition to fuel supply cost advantages (or disadvantages) of a particular location and its energy supply availability or tax liability, the climate at both a broad (macro-) and a local (micro-) scale has a particularly strong influence on building energy use. Climate and its variation is afforded the most attention in this book; the various other, more directly economic or accounting influences on energy use, exerted by selection of a particular building location, are assumed to be more effectively dealt with in books on those topics. The owner is assumed to be capable of influencing the range of options open to the building user for energy management by the selection of a particular location. However, users still exercise a degree of choice in relation to the management of fuel cost or taxation liability.

With climate there is no such option for the building user. Only if the "user" is also the "owner" in the sense employed here can any choice of climate be made. In other words when a user is an owner-occupier or otherwise has some control over the siting of a new, or the selection of an existing, building, that user must by definition also be an owner in our exo-endo model. Dependent upon which aspect of energy consumption is under scrutiny, such a person or organization must examine the owner- or user-related aspects of their behavior in buildings. It is, we feel, one of the strengths of this theoretical model that it identifies, for those interested in energy use in a specific building, which particular aspect to examine first. The owner-occupier can, for example, clarify whether it is a management (user) or building size and location (owner) aspect of energy use that should be examined first in a conservation program.

The owner normally selects a particular size of building on economic and practical grounds related to investment in the site and the extent of the organization(s) to be housed. However, building size is potentially a strong determinant of not only the amount of energy used, but also, more importantly, the intensity of energy use.

It is clear that a large building, housing more people, will consume more energy than a small one because there are more demands to service. What is not so clear, *a priori*, is whether a larger building will consume more or less energy per person or per square meter of floor area than the small one. Data from both theoretical and empirical studies is essential for the owner wishing to make informed selections of building size.

The third and last major endogenous aspect of a building, where an owner's knowledge of the consequences of choices may be effective in framing decisions, is in the activity housed in the building. There is no question that an owner may somehow "decide" to house only activity y or activity z. The issue is the strength of the influence exerted by activity type on energy use in buildings that are otherwise very similar. Awareness of this influence is essential for the building owner interested in target or budget figures for energy use.

3. The Designer

The designer level of the model is next below that of the owner (see Figure 1). At this level designers have to work within a series of constraints imposed by the nation (standards and economic environment), and by the owner (climate, activity, and size). Within these constraints the designer selects a building design that attempts to be an optimum solution of the owner's brief. The designer is constrained to produce a building that not only performs well from an energy consumption point of view but also performs well in all the other roles required of it by the owner and user.

From the selection of construction materials to the choice of heating system, the building designer has a wide range of options. The exo-endo model organizes these only to the extent of labeling the various parts according to their nature in relation to the designer's options. Some building features are neither exogenous nor under the control of the designer; these are the features of a building that are endogenous for the user of the building. The endogenous factors for the building designer must of necessity be organized into one further subset than is implied by the exo-endo split. This is recognized in the structure of the book where two chapters are devoted to designer concerns, one dealing with the energy investment, the other with the energy running costs involved in a building project (Chapters 5 and 6, respectively).

In this context, energy investment is the amount of energy embodied in making and assembling the materials and components of a building. It is an essential part of life-cycle energy costing. To know the "energy cost" of two alternate materials is to possess a further piece of information on which to base a choice between them.

The other, equally essential part of life-cycle costing is the annual cost — how much energy is it going to take to maintain and operate the building services? The second of the two chapters on designers' concerns addresses aspects such as the impact of heating system type, window size, and complexity of services on building energy use.

4. The User

The final level of the exo-endo model presented here is that of the user. All the national, owner, and designer decisions about a building act as constraints on the users' choices. The only endogenous factors left to the user are ones of management. Energy management is essentially a matter of information. Chapter 7 is devoted to the consideration of building users' concerns and the areas of choice, the endogenous factors, open to them. It considers in detail the methods of collecting, analyzing, and organizing information on the energy performance of buildings and groups of buildings. It limits itself, however, to a study of energy management, considering users in a collective manner, rather than as individuals.

It is seen as essential for building users to maintain a close surveillance on the energy consumption consequences of their actions. Chapter 7 does not consider the relative magnitudes of the effects of various decisions on energy use. Instead, it presents a range of techniques for producing such information on the real world performance of one's own building. "Cookbooks" of procedures for energy auditing, retrofitting, or other potentially conserving activities are already available. The unique contribution of Chapter 7 is at the decision-making stage: deciding where to put effort and employ scarce resources most effectively.

The user level is the last level of concern in the exo-endo classification of building energy performance applied in this book. Like all theoretical models not all real situations fit exactly. However, it is our belief that it clarifies more than it obscures and that it identifies those people who may be interested in a particular aspect of building energy performance as well as those who may be able to effect a change in it. Finally, and most importantly, by classifying factors according to their potential interaction with the people involved with buildings, attention is focused where it belongs, namely, on the people and not on the buildings.

III. INDEXES OF ENERGY PERFORMANCE

An index of energy performance is a measuring tool which makes it possible to compare two different levels of energy use in the provision of a particular type of service. It is usually obtained by dividing the energy use by one or more normalizing factors. For example, in transport, the number of liters of liquid fuel consumed per hundred kilometers traveled (or mi/gal) provides a ready index of performance. Such a measure suffices equally for large diesel trucks or small petrol-driven cars. Inevitably some measure of the service provided becomes important; the tonne-kilometer (tonne-mile) measure of freight-carrying ability, for example.

Ever since energy use in buildings became a topic of concern, the search has been on to find a measure of their energy performance. The aim of the search has been to find an index that facilitates comparisons of buildings in the same way that the liters per hundred kilometers (mi/gal) measure does for transport vehicles. Simplicity and general applicability are the key to the success of such indexes. In this section, we shall consider the many lessons that can be drawn from the use and application of transport performance indexes.

A. Energy Performance Vs. Service

The first lesson is the separation of the energy performance from consideration of the service performed. This operation greatly clarifies and simplifies the construction and application of indexes. The operation results from the need to separate building-related energy consumption from activity-related process energy consumption, in order to produce a useful measure of the energy performance of the buildings.

Consider, for example, a long-distance truck hauling a frozen goods container. If one wished to compare the fuel economy of this truck with that of one hauling, say, bales of wool, then the freezer weight would have to be included in the useful load of the truck. The energy used by the freezer would be subtracted from the total energy used by the truck to give the "transport" energy use. The transport performance of the frozen goods truck could then be realistically compared with one hauling bales of wool over a similar route. The process load (freezing) must be removed from consideration of the efficiency of the primary function, that of transport.

If we look now at a building we see similar factors in operation. The energy consumed by processes such as computing or cooking which are housed in a building should not enter into a calculation of energy performance of that building. In practice such processes can often affect the performance of an individual building by altering the operation of one of its energy consuming services (e.g., extra cooling to remove heat gains due to computers). Still, the energy used by the process itself should be considered separately from any assessment of the energy performance of the building. As with the truck it is the energy performance of the primary function (in this case providing the right conditions for the activity) that is of most interest.

There are activity-determined methods of operating or of providing services in buildings that will require buildings to use energy at a variety of rates. Some, like the truck-driver's radio, for example, are part of the overall energy performance measure. In fact, it is often these activity-based differences that are examined during conservation campaigns. Faced with an existing building, little capital to alter it, and instructions to improve energy performance, the person charged with energy management inevitably scrutinizes the amounts of energy used to provide all the services in the building.

The process energy use of the activities housed in the building corresponds to the freezer energy use for the truck. The amount of energy used to cook food or to print books inside a building is immaterial to the calculation of the energy performance index of that building. Each amount is important in the calculation, say, of a restaurant's overall energy efficiency; but, the overall restaurant efficiency index is obtained from the sum of the cooking, food storage, and building services energy consumption indexes.

Hence the first lesson to be drawn from indexes of vehicle energy performance is the necessity to clearly separate the primary activity for which energy is being used (transport) from the secondary energy-consuming activities (freezing in the above example). In buildings, this implies that the primary energy use activity is the provision for the comfort and task-oriented needs of the people housed. Activity-related energy consumption is, as far as possible, excluded from such considerations. The question of the quality of service provided is also avoided; the basic energy performance index assumes all buildings (vehicles) are providing an adequate service. For example, the quality of the thermal environment in the building, like the freight-carrying ability of a vehicle, is assumed to be adequate for the needs of the people using it.

B. Standardization

The next lesson to be drawn from vehicle performance indexes is that of standardization: the "distance traveled" part of the index. That car x uses 10 ℓ and car y uses 5 ℓ is a piece of information that is of little use unless we know that they both used this amount of fuel while traveling the same distance, at the same speed. If we were to be told that while the speeds were the same, car x traveled twice as far as car y in consuming its 10 ℓ we would immediately impose our own standardization. We would calculate how much car x would have used traveling just the same distance as car y. We have no reason to believe that car x consumed fuel at different rates over different parts of its journey so we conclude that it would have consumed half its 10-ℓ total in traveling half the distance. We would therefore conclude that the two cars performed equally according to our standardized index of energy (fuel) performance. Having established a baseline against which the two cars are equal we can go on to address questions of the quality of service provided. For the particular energy performance noted we compare the carrying capacity, comfort, or other "service" performance of the two cars.

With buildings we seek a standardized measure of building performance as well. An index resulting from the standard operation of a building, as for vehicles, permits meaningful comparisons of performance to be made. Again, given such a baseline, it is possible to proceed to further examination of the provision of amenities of the building.

What then can provide as simple a standard for building operation as does travel distance for the transport energy performance index? The setting of a number of standard operating conditions relating to weekly hours of use and internal temperatures is akin more to the specification of a running speed or mode of driving than to the standard distance. Unlike vehicles, buildings cannot even theoretically be required to run for a standard time at a standard rate of energy consumption. External factors like climate preclude such an operation.

For example, depending on the time of year and hence coldness, the energy expended to meet the thermal requirements of people in a building must vary. In the same way that the energy use for acceleration and deceleration and for climbing and descending hills in a vehicle is assumed to be averaged out in practical tests of vehicle energy performance, so the seasonal variation in energy demand in a building must be allowed for by the standardizing process.

The method developed for standardizing building energy performance indexes is to use "annual" energy consumption figures as measures of performance. That figures should be compared over a standard length of time is obvious. The varying seasonal effects on energy consumption can be reduced in overall influence if the standard time always includes the same seasons. The simplest way to achieve this requirement is to include all four seasons in the standard time.

C. Size

Having established an index (megajoules or Btus per year) which can be used to compare energy use in buildings in a way that parallels vehicle fuel use comparisons, the analogy can be extended. As with vehicles, it is possible to consider systematic methods for making allowance for the effectiveness of the provision of amentities. We feel that the extension of the analogy provides useful insights into the process of index creation.

For example, it has become standard practice to see megajoules per square meter per year (Btus per square foot per year) as the index of building energy performance. This is because it is seen as having obvious similarities to the vehicle performance indexes of liters per hundred kilometers (mi/gal). The introduction of the area measure is done to "standardize for size". Dividing annual energy use by building floor area is supposed to make it possible to compare large and small buildings on equal terms. From the discussion in the earlier paragraphs of this section it should be clear that including building area, in the construction of a building energy performance index, is like including vehicle carrying capacity in a vehicle energy performance index. It is an attempt to allow for the service or amenity provided.

While there is more theoretical justification for dividing annual energy use in buildings by some measure of building size, it is not so clear which measure is the most suitable. The measures of size are many; they range from the area and volume of the building through its hours of operation to the number of people it houses. All can be expected to have a relatively "linear" relationship with energy use according to any of the following rationales:

- Two times the physical size requires twice the heating energy
- Two times the current hours of use requires twice the operating energy
- Two times the number of people require twice the services

assuming the quality of the environment provided remains similar.

The most commonly used "size-normalizing" factor is area. Whether based on net, gross, or conditioned area, the Area Energy Use Index (AEUI) is the index most quoted in standards,[5-7] auditing manuals,[8] or other documents[9] examining and comparing building energy performance. It is a very easy measure to understand being "just" the annual energy use per square meter (foot) of a particular building; the total annual energy use divided by the net, gross, or conditioned area of the building. It has considerable appeal because it possesses many parallels in the everyday world: watts per square meter (foot) for lighting or dollars per square meter (foot) for leasing commercial premises. This ability to be widely applied or compared is one of the sought after features of an energy performance index; an index is more likely to be used or calculated, if the information needed is already relatively familiar.

There are many limitations to the AEUI and these have been pointed out by various researchers.[10,11] The limitations are related principally to the many variations in the use of buildings that are not allowed for by a simple area normalization. These limitations are that division by area may be only partially successful in removing size effects and in some cases may not even account for the different uses to which people in the building put the energy. It may not be relevant to restaurant owners, for example, that their restaurant has a higher than average AEUI; they will however be extremely interested in energy use per table or per customer.

Other size-normalizing factors have been investigated and have produced interesting results.[12-14] The number of people in, and the volume of, a building have both proven at least the equal of area as size-normalizing factors, and have in some cases been found to be more effective. Hours of use (by comparison with a "standard" of 50 hr/week) has proven a useful supplementary normalizing factor.

D. Activity

It is interesting to note that the approach often adopted with vehicles, that of comparing like-with-like, is not so often taken with buildings. A car and a truck are seldom compared in vehicular performance analyses; this is because it is hard to think of normalizing factor(s) with which to standardize the performance indexes so as to make useful comparisons. Comparing the building equivalents of "car" and "truck" instead of "car" and "car" is often done in building performance analysis, hence the search for further normalizing factors.

The only parallel that can be drawn for buildings, in this like-with-like comparison, is the strong tendency, in building energy performance analysis, to label buildings according to the activities they house. Thus, "school" buildings and "office" buildings are normally studied separately. However, for any given activity, there will be a wide range of building sizes and types to be compared. A performance index is needed for the comparison of different sizes and types of buildings housing different activities.

There have been several attempts to generate more complex indexes of building energy performance making use of multiple normalizing factors.[15-17] No doubt many more will follow as researchers and building operators try to come to grips with the inadequacies they find in using the AEUI as a performance index.

Within a given activity it is often possible to produce a better performance index than AEUI. For office buildings with largely clerical functions the energy use per person (PEUI) is a more useful measure than AEUI. However, in general, multiuse buildings the AEUI is almost as good a measure as PEUI since there is a very strong relationship between the number of occupants and the floor area, that is, the density of occupancy does not vary greatly.

E. Other Factors

The three other factors frequently used in the construction of building energy performance indexes are hours of use, severity of climate, and the type of activity housed in the building. Climatic severity and hours of use are best allowed for by dividing annual energy use per unit area by a factor that is constructed on the basis of some norm of climate or duration of use. Thus a climate that is 5% colder would be expected to have a "heating" energy use index that is approximately 5% greater. To construct such an index properly, the proportion of the total energy used for heating must be determined or estimated so that the normalizing process is applied only to the correct portion of the total energy use. The last of these three factors is best allowed for by treating each activity as a separate category, with a separate index, as indicated earlier.

F. Summary

It is apparent that there is no one universal index of building energy performance and it is unlikely that there ever will be one. Indexes are chosen, not because of the particular properties or behavior of the building or of the people in it, but to be used in a particular analysis. While this may be a truism it is still not widely enough recognized in practice since, too often, indexes are misapplied because they have not been constructed with the particular application in mind.

The most obvious example is in energy conservation programs. The manager trying to decide in which of a number of buildings to start an energy conservation program does not wish to know which building uses the most energy per square meter of floor area. That manager, as discussed in more detail in Chapter 7, wishes to know what program of action, for the least expenditure of resources, will have the greatest return in money rather than in energy terms. The same manager also wishes to make allowance for the fact that despite using more energy, some buildings are more effective housers of activities, in the sense that they contain people in greater densities. A performance index, which permits a ranking of buildings on one scale according to these criteria, is better for the manager of a number of buildings. A simple "one factor" annual AEUI or PEUI measure of performance is unlikely to be appropriate.

For other purposes, such as quick comparisons of energy use rates of alternative premises for lease, AEUI may be useful; though even for standards, where simplicity is important, AEUI has its limitations. The CIBS Building Energy Code,[6] for example, uses a number of multipliers which can be used to convert base AEUI figures into an index that allows for individual variations in conditioning system type and other factors. Without such multipliers the energy targets are of little use to building operators wishing to make comparisons with the actual annual energy use of their own buildings.

Because external factors like climate can have a more identifiable and more marked effect on energy use in buildings than their parallels do on energy use in vehicles, the normalizing factors that can be applied to the basic annual energy use measure (index) of performance for buildings are more numerous. Often it is more a case of "must" rather than "can", because particular applications must use particular normalization factors. To allow comparison or analysis of two apparently dissimilar buildings, we must have particular normalizing factors applied to their energy performance indexes to remove the disparity which is not relevant to the analysis. If the normalization is successful, the disparity that is relevant is highlighted on a simple scale, that is, an index of performance.

And finally we have the most obvious lesson of all from our vehicle analogy. The more widely understood and applicable a standardizing (normalizing) factor is, the more the standardized index will be used. For example, if it is desired to readily compare energy use, rent, lighting loads, investment in office equipment, and building population, then the AEUI must be used because it is simple and easy to understand and the process of generating floor area based indexes like it is familiar.

IV. APPLICATIONS OF ENERGY PERFORMANCE INDEXES

The preceding section of this chapter introduced the idea that the applications required of a performance index should be the prime determinant of its form. This section expands on the form of index used in a range of applications. The concerns of the four interest groups defined in the early part of this chapter are each addressed in this section. There are four parts.

The first part, following these introductory paragraphs, considers energy use indexes of interest to the nation and to the building owners of the nation. These are the concerns of Chapters 3 and 4 of this book.

The second part explains the indexes of interest to the designer who is examining the amount of energy to be invested in the materials and construction of a particular building. Like Chapter 5, the concern is the energy embodied in a building or its constituent components and materials.

The third part is devoted to another concern of the designer, energy consumption during the day-to-day running of the building. For the designer of new or upgraded buildings this is usually a predicted value based on some physical model of energy use in buildings. Chapter 6 considers these concerns in detail.

The establishment of energy performance indexes using data on actual energy use in buildings is the subject of the final part. This collection and analysis of real data is considered to be primarily the concern of the people who operate and use the building, its management, its system operators, and its occupants. Chapter 7 expands on this topic placing special emphasis on management aspects.

A. National and Owner Concerns

The nation and the individual building owner are considered together in our examination of performance index applications. This is because of the considerable influence that the nation can have on the owner's behavior and attitudes. There is a strong temptation for governments wishing to lower the national building energy demand to set some type of energy performance target for the building owners of their nation. Mandatory design or in-use performance targets should affect the owners' energy-related action and they should also affect the type of building an owner might consider building or buying.

The principal form of index applied to date by nations has been annual energy use per square meter (foot) of building floor area (AEUI). This is an index that is both simple to calculate and simple to understand, but not so simple to apply or enforce. In the main the enforcement problems arise because the index is so general. If it is mandated that typical retail premises must use less than a particular AEUI value who is to measure or test how typical a particular building is? Who can objectively measure such a feature? The ease with which total energy of a building can be evenly spread throughout that building, through the use of indexes like AEUIs, is attractive to governments and building owners because such calculations are routinely performed by both groups.

The standardizing factor most commonly used is area. Using quantity x per square meter (foot) rather than the total of x for the building is supposed to make possible direct comparisons between buildings; similarly, rental charges are calculated per square meter (foot) to better enable comparisons between buildings. However, it is unwise to apply simple across-the-board AEUI figures to a group of buildings. In the same way that a building owner will look at what each dollar per square meter (foot) will buy in additional services in each building, so a building owner should examine not only the AEUI but also the services provided by the building. Prospective owners buying a building, or having one constructed, should have an AEUI for the building calculated even though this is not the final step. Service or amenity provision must also be compared and minimum AEUI for maximum service should be the aim. The juggling of these two ideas permits the ready identification of trade-offs between lowered running cost and lowered amenity.

The methods of calculation of AEUI are many and varied due to the wide range of possible ways of estimating both building area and energy use. Building area can be estimated on a gross, net, or other more complex basis, a matter of definition. The differences for energy use are more deeply rooted. In the design stages of construction of a building, the total energy use can only be estimated, if necessary by a very sophisticated computer simulation. Ideally, the total energy use aspect of an AEUI calculation should result from measurements obtained in well-managed buildings that have been commissioned and are well operated. Only in this situation can the many interactions between building, HVAC system, and building occupants be accounted for. The same restrictions or limitations apply for nations as apply for building owners. Calculations based on "typical" buildings which are "operated as designed" do not agree with real world information. Running such a "prototypical" building through a number of computer simulations cannot produce particularly useful information. This is because the all-important user-behavior equations have not yet been solved; in fact they have not yet been formulated. Even more than the individual building owner, the nation, through its government, has more reason to be interested in these user influences. It is at the national level in large groups of buildings that the greatest effects of building occupant behavior will be noticed. To ignore this behavior is to tempt fate in the application of an energy performance index.

There are many indexes, more complex than the AEUI, that could assist building owners, if not nations, in decision making about building energy use. The inclusion of measures of productivity, the allowance for HVAC system efficiency, or the application of various activity factors could all be part of the generation of an individual owner's building energy performance index. However, these considerations are most suited to the individual building owners' management calculations because the circumstances of each building are unique. A total performance index can only be generated in this detailed way specifically for individual buildings. The formulation of a general, but detailed performance index, which measures service as well as energy performance, ought not to be attempted. But a general performance index, that forms an initial energy use baseline in terms similar to those formulated for other building service performance measures, is required. The AEUI, or its equivalents which use building volume or building occupant numbers to standardize the energy use, are essential starting points. The AEUI has the edge in buildings or for activities where the other productivity or performance measures are conventionally standardized by dividing their total values by building area. Being able to compare two indexes standardized in the same way is both a simple and a useful task. Because of this, the AEUI is likely to become the most common index of building energy performance, despite the shortcomings previously outlined. It is apparent that at all levels of the exo-endo classification system the AEUI is the most widely applicable, and hence the most readily applied, measure of energy performance.

The nation, in the person(s) of its government, influences building energy performance both directly and indirectly. As will be explained in more detail in Chapter 3, this is because the nation is concerned both with the energy used by its citizens and with their health and safety. In fact these latter criteria are the more usual concerns of standards generated out of national interests because the specification of environmental quality performance criteria is the business of health and safety standards. Often these requirements make it unavoidable that occupants of a building use its energy-consuming services more than they might otherwise have chosen.

The introduction of energy performance standards has forced a reappraisal of the bases of the healthy and safety standards in force in many countries. In lighting for example the idea that "more is better" is being tempered by a realization of the energy cost of providing more. Quantification of the real benefits, to trade-off against the actual energy cost, is now sought. It is as true for national standards as it is for individual building owners, however, that these other benefits of energy consumption ought to be considered independent of any energy performance index. A generalized index such as AEUI will be useful in any trade-off process, but the process ought to consist of a series of guidelines in a standard whereby the building owner is shown how trade-offs might be made for their own individual situation. The single, unique, all-purpose performance index for energy use and amenity provision has yet to be devised. It is unlikely ever to be devised as it would be too complex to be useful except in a purely theoretical model. Inclusion of all the necessary qualifying factors, needed for such an index to be widely applicable, would inevitably divorce it from the practicalities of application. Once again, the AEUI comes to mind. Its attraction is its simplicity. It is widely applied, though not in an ideal sense widely applicable; it does not take into account the many influences on energy consumption which can vary tremendously from building to building. It is, however, simple to calculate from readily gathered data.

Hence, the AEUI, rather than a more complex and potentially more widely applicable index, is likely to continue to find wide application in practical situations.

B. Designer Concerns — Capital Energy Costs

To consider here all the ways in which the information about the energy cost of materials might be indexed is to needlessly duplicate the material contained in Chapter 5. The principal concern of this section is the presentation of a broad overview of the appropriate indexes, their calculation, and their applications.

The first step in any consideration of "embodied energy", the investment of energy required to produce a component or service, is the definition of a few terms. While these will be explained in detail in Chapter 5 it is useful to understand the difference between Gross and Process Energy Requirement (GER and PER) at this stage.

The GER of a building or of a component in that building, is comprised of the sum of:

- The energy input into the support services and transport at the building site
- The energy input into the construction (or manufacture) of the building (or component)
- The energy consumed in extracting, processing, and transporting the original raw materials for the building (or component)
- The energy used to construct the plant used in the extraction and processing of the raw material
- The energy cost of repairing damage caused by the component manufacturing process

The PER is a very restricted subset of the GER. The PER for a component is the energy used in its manufacture. It might, for example, be the energy input per tonne output of a nail factory. No detailed study is made of raw material extraction or environmental impact in a PER calculation. Typically, the PER ranges from 50 to 80% of the GER for any one product.

There are many ways of looking at constructing an index for embodied energy performance. These ways are based yet again on standardization. The presentation of research figures for embodied energy is most often per unit of material. Thus cement, for example, has a PER of 8.05 MJ/kg (3461 Btu/lb). This is a very restricted format for such information. One is often more interested in how much energy is required to construct a concrete frame ten-story building where cement is the essential ingredient. Similarly the publication of a figure of 35.3 MJ/kg (15,175 Btu/lb) for steel nails makes little sense to the designers of timber-frame dwellings. They need to know the significance of this to the embodied energy total for a whole dwelling. Examples of both formats for presenting such information are given in Chapter 5, along with a proposal for a computer program for calculating the one from the other.

In a parallel way, the imported vs. indigenous energy use balance of building products is of limited value. It is only of interest where a government is concerned about the amount of energy imported in processed products which could be substituted by processing locally using indigenous energy. Individual designers, in most cases, will find such extra considerations irrelevant to their building projects. International trade and politics, while influencing building design, are such complex topics that attempting to include them as design parameters is not desirable. A building must be designed within the current, largely exogenous, political, and economic conditions. It is to be hoped that these conditions are the result of rational resource allocation decisions so that the designer's own decisions (in reaction to them) are correct from a national viewpoint.

Another method of standardizing the embodied energy in building construction is to set up figures for elements of a standard building. In this method, the energy use figures for the roof, walls, windows, and floor of, say, a standard house would be available. Using this approach, the establishment of comparative figures for different materials which achieve the same purpose becomes possible. For example, alternative roofing materials with similar durability and thermal resistance can be compared on the basis of their respective energy investment. As always, the difficulty is one of defining what is standard. A standard house or a standard office building is something about which it is hard to obtain general agreement.

For practical purposes it is apparent that this standard building approach is most appropriate to the designer when the building is at the initial concept stage. It provides readily assimilated information that can be developed by the individual designer into guidance for the task in hand. The problem of the definition of a standard can be overcome by careful description of the basic building. At later design stages, the more detailed approach becomes more useful. It is anticipated that this detailed capital ''energy'' cost calculation will parallel the capital ''money'' cost approach of the quantity surveyor. A ''bill of energy investment'' could be produced at the construction documents analysis stage.

In conclusion, there seems wide agreement that the best way to measure the significance of the embodied energy index of a building is to compare it with the annual operating energy requirement. Thus, the number of years of normal operation it takes to reach the embodied energy total, becomes the desired index of energy investment. Life-cycle costing, where the lifetime energy use and the initial energy investments are considered in similar (that is inflation-adjusted) terms, is the ideal approach.

C. Designer Concerns — Systems Energy Performance

The indexes of use to the designer are basically the same as those which concerned the owner and the nation. The AEUI is the most often quoted index in this context. Yet again, this choice is dictated by the familiarity of the area standardization procedure rather than a rational selection of the most appropriate index. There is little that can be added at this point that has not already been said for owners or will not be covered in Chapter 6. The principal points to be remembered about energy consumption performance indexes generated for designers are

- There is likely to be a considerable disparity between reality and the predicted performance of a new building design. This disparity is generated by the in-built assumptions in the simulation and by the assumptions made by the people entering the building description data to the model.
- In practice, much of the theoretically derived dependence of building energy performance indexes on building features can be swamped by effects resulting from the activities and behavior of the people who occupy the building.
- As with other uses of energy performance indexes, the simplest method of incorporating the many other, often conflicting, performance criteria set by a building design, is to deal with the energy use figures in the simplest possible way. An AEUI is sufficient for most purposes. The complexities and trade-offs inherent in design can be performed by using AEUI as a ranking parameter during the consideration of other widely different aspects of performance. One avoids creating an ever more complex index, based on measures of the quality of the services provided by the use of energy in a building; instead, various values of AEUI are identified as offering different options for the performance of these services. For a particular AEUI value there is then a corresponding service option.
- To a large extent the alternative standardizing factors can best be incorporated in the above fashion rather than as extra parameters in the calculation of a complex index. Even the obvious adjustments to AEUI, like climate dependence, are more manageable when externally applied to, rather than incorporated in, an index. Rather than state AEUI per (temperature) degree, it is far easier and more widely understood, to present various AEUI values corresponding to different climates.

The building designer has many reasons for standardizing building energy use data. All the reasons relate, again, to making comparisons. They range from comparing the current project with the previous one to examining differences in energy use between various design choices.

It is important to consider carefully the information sought in any comparison, as this may govern the standardizing factor selected. The simplest index, suitably modified by, say, categorization of the parameter being investigated, is the index that is most readily understood. Hence it often produces the most useful information. Designers trying to balance various criteria within familiar terminology are more likely to understand the consequences of their choices than those who feel constrained by a complex and abstrusely formulated energy index.

D. User Concerns — Management of Building Energy Consumption

For this last category of concerns the simple AEUI yet again forms a useful base from which to make some comparisons. The comparisons are just as frequently between energy use records for the same building during different time periods as they are between records for different buildings altogether. The former type of comparison occurs when a check is being kept on the operation of a building. AEUI is largely irrelevant to this process as long as the areas are unchanged. Different buildings are compared by owners when they seek an absolute rather than a relative measure of performance for their particular building. Comparison of the current energy performance of a building with its past record should happen on a regular basis in any well-managed building if an adequate check is to be kept on operating efficiency. Such a comparison should also form an integral part of an energy audit or a conservation program. In this situation its purpose is to look for real changes in building performance. Allowance must be made for differences in climate within the index. Area normalization is unnecessary unless radical alterations have been made to the building itself. Climate normalization, at least by a measure of the coldness of the climate, is essential. Measures such as the PSA Performance Line,[18] or division of the heating portion of the energy use by a degree-day-based measure of coldness, have been attempted. Both have had some measure of success. The performance line approach provides the more suitable index of performance for the building user or occupant.

The performance line approach compares energy use for a particular period against the average temperature for that period. Both these values are simple to obtain. The energy use figure usually comes from energy suppliers' records or bills; the period is the suppliers' billing period. The average temperature can be obtained easily from the nearest meteorological office. As long as the meteorological data available is systematically related to the climate of the site, actual temperature data for the site is unnecessary. What is of interest in the performance line method is the variation from year to year in the slope of a line on a graph of consumption vs. temperature. The concept is explained more fully in Chapter 7 but the basic ideas are clear; the index is simple to generate, it can be continuously updated with the latest performance data making it an ideal system management and control monitor, and it graphically highlights differences from year to year in the operation of a building.

It is as important for building occupants to compare energy performance for different buildings as it is for building designers and owners. The principal motivation for doing so is often financial. It is the occupants who, directly or indirectly, pay the energy bills.

In addition to establishing a baseline for measuring the performance of a particular building, building users may wish to compare building performance when selecting new premises or deciding on a conservation program. Chapter 7 contains suggestions about indexes for all these cases. As is to be expected from the discussion earlier in this chapter, each purpose requires a separate index. To establish whether a building is a high or low consumer requires no more than an AEUI measure of performance. Establishing rational criteria for the investment of money on a conservation program requires an index that combines AEUI or some other measure of building energy efficiency with the present cost of energy. Such an "index of energy conservation potential" is described in Chapter 7.

V. ASSESSMENT OF ENERGY PERFORMANCE

Estimation of the energy performance of a building, as has been explained in this chapter, is a process to be undertaken with some care. The concept has utility in many different spheres, and careful attention must be paid to selecting the measuring tools to be used. These tools consist of a series of indexes which differ both in formulation and function. The selection of an index for assessing energy performance in buildings must take account of both the activity which uses the energy and the reason for assessing the performance. Simple area-based indexes are useful, for example, where one wishes to compare energy use per unit floor area with the area-based costs of other services provided in the building.

The exo-endo classification scheme presented at the beginning of this chapter (see Figure 1) assists in the identification of the type of index to be used in a particular assessment of building energy performance. It also identifies clearly the organization "concerned" in the assessment. The whole structure of this book has been built around this exo-endo classification. It identifies four primary levels of concern at which particular groups are usually able to make decisions which affect energy use in buildings.

The first level is that of the nation. The choices open to a nation, especially in the area of targets, are described in detail in Chapter 3. The nation has its exogenous factors (constraints) generated by international factors. All the other levels of the exo-endo classification have constraints that are generated by the nation. The decisions made by the owner among the endogenous factors (choices) at the next level of the classification add to the exogenous factors affecting the last two levels. This process of ever-increasing constraints and ever-decreasing choices continues on through to the designer and the user levels of the hierarchy.

Within each level of the exo-endo classification, different measures of energy performance are applicable. We shall now look in more detail at national concerns.

REFERENCES

1. **Markus, T. A. and Morris, E. N.,** *Buildings, Climate and Energy,* Pitman, London, 1980.
2. **Hillier, B. and Leaman, A.,** A new approach to architectural research, *RIBA J.,* p. 517, December 1972.
3. **Olgyay, V.,** *Design with Climate — A Bioclimatic Approach to Architectural Regionalism,* Princeton University Press, 1963.
4. **Garmonsway, G. W. and Simpson, J.,** *The Penguin English Dictionary,* Penguin Books, Harmondsworth, 1965.
5. Energy Performance standards for new buildings: proposed rulemaking and public hearings, *Fed. Reg.,* 44, 230, 68120, November 28, 1979.
6. CIBS Building Energy Code, Chartered Institution of Building Services, London, Part 1, 1977; Part 2, Section (a), 1981; Part 3, 1979; Part 4, 1982.
7. NZS 4220: 1982 Code of Practice for Energy Conservation in Non-Residential Buildings, Standards Association of New Zealand, Wellington, 1982.
8. **Dubin, F. S., Mindell, H. L., and Bloome, S.,** How to Save Energy and Cut Costs in Existing Industrial and Commercial Buildings, Noyes Data Corporation, N.J., 1976.
9. **McCarthy, P. M., Patel, R. F., and Karpoy, B.,** Empirical and Simulation Analysis of Energy Use in Commercial Buildings, HIT-664-2, Hittman Associates, Columbia, Md., February 1977.
10. **Spielvogel, L. G., Orlando, J. A., and Hayes, P.,** Feasibility of an energy index for office buildings, *ASHRAE Trans.,* 84 (Part 1) 1978.

11. **Donn, M. R.,** The Feasibility of Energy Use Indices, Proc. 1979 Annu. Conf. New Zealand Institution of Engineers, Wellington, February 1979.

12. Enviro-Management and Research, Evaluation of Building Characteristics Relative to Energy Consumption in Office Buildings, Washington, D.C., September 1975.

13. **Baird, G., Donn, M. R., and Pool, F.,** Energy Demand in the Wellington Central Business District — Final Report, Report 77, New Zealand Energy Research and Development Committee, Auckland, 1982.

14. Syska and Hennessy and Tishman Research Corporation, Energy Conservation in Existing Office Buildings, Phase 1 Report, Department of Energy, Washington, D.C., June 1977.

15. **Anderson, T.,** The energy man, *Building Design,* April 18, 1980.

16. **Baird, G. and Brander, W. D. S.,** Energy Conservation in Government Buildings — Report on Stage 2, Contract Research Paper 5, School of Architecture, Victoria University of Wellington, New Zealand, February 1982.

17. **Baird, G. and Chan, S. A.,** Energy Cost of Houses and Light Construction Buildings and Remodelling of Existing Houses, Report No. 76, New Zealand Energy Research and Development Committee, Auckland, 1983.

18. PSA, Monitoring Fuel Consumption of Heating Installations, Technical Instruction (M&E), M79, Department of the Environment, Property Services Agency, Croydon, U.K., December 1974.

Chapter 3

NATIONAL CONCERNS — INSTITUTIONAL ROLES AND ENERGY STANDARDS

I. INTRODUCTION

In the previous chapter, a detailed description was presented of a system for classifying the factors that influence the energy performance of buildings. A descending hierarchy of levels of concern was identified, ranging from national concerns at one extreme to the concerns of the building user at the other.

In this chapter, the concerns of major national institutions — governmental, professional, industrial, etc. — are identified, and the most important manifestation of these national concerns from our point of view — the promulgation of building energy performance standards — is reviewed in some detail.

II. THE INTERNATIONAL SETTING

We start at the broad end of the spectrum of the exogenous-endogenous (exo-endo) classification system by examining the international setting with respect to building performance. This setting provides the exogenous constraints to national institutional concerns. Government policies, professional requirements, and building industry activities must all respond to international factors.

The basic constraints governing the current global energy supply and demand situation are well documented elsewhere[1,2] and will not be explored in detail here. No one reading this book will need to be apprised of the energy situation of the world. It has been reviewed many times from a variety of viewpoints, particularly since 1973. Even before that time, the constraints of the finite nature of the resources of the world had been well argued,[3] indeed predicted.[4] Of course, it was possible at that time to dismiss such arguments as the interesting, but purely theoretical conclusions of the ''environmentalist'' movement.

Much less easy to ignore — impossible in the case of the nations of the world with industrialized economies — were the practical constraints of global policies as manifested in the actions of the members of OPEC. These industrialized countries were dramatically forced into a reappraisal of their own national energy situations. In many cases this was probably the first time that an appraisal of any depth had been undertaken, certainly since wartime, to judge from the upsurge in the formation of energy ministries and research and development agencies.

Much of the initial activity of these organizations related to problems of energy supply and demand. The energy supply side of the picture proved the easier to assess, and results followed relatively quickly. Assessment of energy demand took rather longer, and it was some time before the picture began to emerge. When it did, it was apparent that a significant proportion of the energy use of industrialized countries was related to the provision of heating, cooling, lighting, hot water, and other energy-consuming services for the occupants of commercial and domestic buildings. A 1980 survey[5] of a number of developed nations indicated that the commercial and domestic sectors together consume, on average, 37% of the total primary energy used. There was also evidence of considerable potential for reductions in energy use, without any deterioration in conditions inside the buildings. In such a situation it is not surprising that standards of building energy performance are receiving increased attention.

A. Building Energy Performance Standards — International Trends

Most developed countries have standards (or codes or regulations) which deal with one or more aspects of the thermal performance of new buildings. Thermal insulation values for walls, roofs, and floors are specified, virtually without exception, either as desirable targets or, increasingly, as mandatory requirements.

Other aspects of energy use in buildings are also being made the subject of new standards, particularly in the Scandinavian countries. In Norway, for example, upper limits are placed on the average thermal transmittance of the building facade (wall plus window), but an allowance is made for the effect of any solar heat gains via the windows.[6] The airtightness of a building is now the subject of standards in both Norway and Sweden. Regulations relating to different aspects of the design and operation of some of the energy-consuming systems of buildings have been introduced in Denmark, Finland, and Sweden. These include stringent, energy-related criteria for the design and commissioning of central heating and air-conditioning systems,[7] indoor air temperature limitations and requirements for the provision of controls to regulate energy use,[8] requirements for operating and maintenance instructions, specification of the conditions under which exhaust air heat recovery must be undertaken, and the type of provision to be made for the measurement of energy use.[9] Other countries are also moving in these directions. Increasing attention is being given to such matters as the control of ventilation and infiltration losses, the more effective use of solar and other heat gains, and improvements in the efficiency of the energy-consuming systems themselves.

Most of these measures deal with specific individual aspects of building design. However, several countries have made significant strides towards regulations which deal with the energy performance of the building as a whole. These range from the Irish standard which regulates the overall heat loss of a domestic building according to a volume to surface area ratio, to the state of California requirement that new nonresidential buildings be within an allotted energy consumption budget figure, given specific standardized design conditions. Much of this chapter is devoted to an exploration of this trend towards energy standards which treat the building as a whole. While most of these "whole building" standards have been concerned with new buildings, existing buildings have not entirely escaped scrutiny. For example, Sweden has set a target for a 30% reduction in energy consumption in the building sector for the 1978 to 1988 decade.[9] In New Zealand, a provisional code of practice has been introduced which specifies energy targets for existing buildings (more of this later).

B. The Role of International Agencies

In the face of a common problem (in this case, energy) it comes as no surprise that an increasing number of international programs, exchanges, and other cooperative ventures have been undertaken. Some of these ventures, like the formation of the International Energy Agency, involve the direct commitment and cooperation of governments. Others involve the development of new programs within existing international institutions, governmental, professional, and industry based. The plethora of international conferences and symposia on energy and its use in buildings provides ample evidence of a mounting concern in this field, a concern which transcends national boundaries.

Government agencies, professional and research organizations, and industry groups have been quick to see the advantages of such international cooperation. What follows is an outline of the broad scope of two of these programs, together with a brief mention of some other important activities.

1. Inter-Governmental Activities

While it cannot be denied that some inter-governmental cooperation on energy matters existed before 1973, most of it related to energy supply, and very little was concerned with energy performance. International agencies such as the United Nations had been involved in the sponsorship of work related to the thermal performance of low cost housing in underdeveloped countries. The results of the work of Koenigsberger et al.[10] represent an important step in the drive to improve the energy performance of low cost domestic buildings.

Since 1973, however, a more concerted attempt has been made, particularly by the industrialized nations, to cooperate on matters of mutual concern in the energy field. One of the principal vehicles for this type of inter-governmental activity is the International Energy Agency (IEA). Set up in 1974 at the request of the (then) U.S. Secretary of State, Dr. Henry A. Kissinger, the IEA is an autonomous body within the Organization for Economic Cooperation and Development (OECD). It is comprised of 21 industrialized countries (see Table 1) which have formally agreed to cooperate on energy policy. Although primarily concerned with security of energy supplies (oil in particular) for the participating nations, the IEA is increasingly looking towards energy conservation and new technologies as a means of reducing dependence on imported fuels by their member countries.

As part of this cooperative effort, the members have agreed, not only to carry out national programs of energy research, development, and demonstration, but also to undertake jointly funded and task shared projects.[11] The purpose of the latter cooperative type of activity is to avoid unnecessary duplication of effort and allow scarce resources to be used more effectively. Three specialist working parties deal with end-use technology, renewable resources, and fossil fuels. The principal task of these working parties is to identify areas for inter-governmental cooperation and to promote specific cooperative programs.

One program, which was developed early, is of direct relevance to the energy performance of buildings. Styled the "Energy Conservation in Buildings and Community Systems Program", it has provided an invaluable international forum for the discussion of energy and building problems and has given the participants a global perspective of the situation.

Several projects have been undertaken under this program. For each project, several countries may be involved, in some cases as few as two, in others ten or more. One country is responsible for the management of the project; all share in the tasks and/or the funding. Table 2 gives a brief outline of all the existing and proposed projects of their 1981 program.[12] Several of them relate specifically to key issues in the prediction and measurement of building energy performance. It is a short step to a direct consideration of energy performance standards.

A further example of inter-governmental cooperation is the extensive program of energy conservation research and development which has been established under the auspices of the Commission of the European Communities. This program deals with a variety of topics, ranging from the energy optimization of industrial processes to the application of heat pumps.[13] Although building energy performance standards are not mentioned per se, several projects concerned with the improvements of building insulation and the control of heating and ventilating systems are clearly relevant to the formulation of such standards.

2. Inter-Professional Activities

The International Council for Building Research Studies and Documentation (CIB) takes possibly the most comprehensive interest, of any international body of this type, in building and the building industry. Its overall aims are to ". . . encourage, facilitate and develop international cooperation in building research and ensure its effective dissemination."[14]

Virtually all the major building research institutes of the world are members of CIB. Nearly 60 countries are represented (see Table 1).[15] The CIB operates through a network of working commissions and steering groups each of which specializes in a particular topic. For example, Steering Group S17 on Heating and Climatization is concerned with research into heating and air-conditioning systems, the energy consumed by such systems being an important item in any consideration of building energy performance. A variety of working

Table 1
NATIONAL MEMBERSHIP OF IEA,
CEC, REHVA, AND CIB

Country	Organizational membership			
	IEA	**CEC**	**REHVA**	**CIB**
Algeria				o
Argentina				o
Australia	o			o
Austria	o			o
Bangladesh				o
Belgium	o	o	o	o
Brazil				o
Burundi				o
Canada	o			o
China				o
Cuba				o
Czechoslovakia			o	o
Denmark	o	o	o	o
Egypt				o
Ethiopia				o
Finland			o	o
France		o	o	o
W. Germany	o	o	o	o
E. Germany				o
Ghana				o
Greece	o	o		o
Guatemala				o
Hungary			o	o
Iceland				o
India				o
Indonesia				o
Iran				o
Iraq				o
Ireland	o	o		o
Israel				o
Italy	o	o	o	o
Japan	o			o
Korea				o
Luxembourg	o	o		o
Mexico				o
Nepal				o
Netherlands	o	o	o	o
New Zealand	o			o
Nigeria				o
Norway	o		o	o
Pakistan				o
Poland			o	o
Portugal	o			o
Romania				o
Singapore				o
South Africa				o
Spain	o		o	o
Sri Lanka				o
Sudan				o
Sweden	o		o	o
Switzerland	o		o	o
Tanzania				o
Togo				o

Table 1 (continued)
NATIONAL MEMBERSHIP OF IEA, CEC, REHVA, AND CIB

	Organizational membership			
Country	IEA	CEC	REHVA	CIB
Turkey	o			o
U.K.	o	o	o	o
U.S.	o		o	o
U.S.S.R.				o
Venezuela				o
Yugoslavia			o	o

Note: IEA, International Energy Agency membership;[11] CEC, Commission of the European Communities; REHVA, Representatives of the European Heating and Ventilating Associations;[17] CIB, International Council for Building Research Studies and Documentation.[15]

Table 2A
IEA ENERGY CONSERVATION IN BUILDINGS AND COMMUNITY SYSTEMS PROGRAM[12]

Current Work

Project Title, operating agent, joint funded/task shared	Outline of project
Energy Conservation Measures for Heating Residential Buildings; Sweden; task shared (six countries)	This project involves three tasks: development of calculation methods to predict energy savings in residential buildings; production of a handbook on the design of experiments, instrumentation, and measuring techniques; evaluation of national case studies and generalization to other countries
Glasgow Commercial Building Monitoring Project; Glasgow University (BSRU); joint funded (six countries)	This project involves the detailed measurement of environmental conditions and energy flows in an office building; the data will be used to check and improve the design of energy prediction computer programs
Air Infiltration Center; Oscar Faber Partnership (U.K.); joint funded (ten countries)	This center has been set up to provide technical support for those researching into air leakage in buildings; its aim is to develop tools which can be used to assess energy losses due to air infiltration
Building and Community Systems Energy Systems and Design of Communities (ENSYDECO); Greece; joint funded (four countries)	This project aims to develop an energy sensitive planning methodology for communities; the first phase involves participating countries in the evaluation of current national projects
Local Government Energy Projects; Department of Energy (U.S.); task shared (four countires)	This project aims to optimize the design of local energy programs by matching available resources with consumer needs; six subtasks are involved

Table 2A (continued)
IEA ENERGY CONSERVATION IN BUILDINGS AND COMMUNITY SYSTEMS PROGRAM[12]

Current Work

Project Title, operating agent, joint funded/task shared	Outline of project
Inhabitants Behavior with regard to Ventilation; Switzerland; task shared (five countires)	The main objectives of this exercise are correlation of occupant behavior, in respect of ventilation, with climate; estimation of resulting energy loss; study of behavior motivation; study the effect of behavior modification educational program
Minimum Ventilation Rates; W. Germany; task shared (six countires)	Main objectives are to propose objective criteria for evaluating ventilation criteria, conduct a background study concerning minimum ventilation standards, prepare an appropriate R & D program
System Simulation; Belgium	Accurate simulation of HVAC system operation is very difficult; the objectives here are to develop improved methods of equipment simulation, provide reliable simulation techniques for use by engineers, establish methods for improving control strategy, improve methods for equipment selection, establish a data bank for HVAC component performance, and provide realistic guidance on assumptions for lighting, occupancy behavior, and other internal gains

Table 2B
IEA ENERGY CONSERVATION IN BUILDINGS AND COMMUNITY SYSTEMS PROGRAM

Proposed Work

Project Title	Outline of Proposal
Energy Auditing; Sweden	The aim of this proposal is the development of cost effective means, methods, and strategies for auditing in buildings. Two subtasks are identified: data collecting methods and instruments for auditing and identification of appropriate retrofit packages
Windows and Fenestration; Netherlands	Basically, a review of existing window systems, window physics, and calculation methods, together with sensitivity studies of the effect of window, building, climate, and occupant behavior characteristics
Thermal Energy Metering; U.S.	The main objective is to evaluate different fluid metering concepts with a view to the development of a low cost thermal energy meter
Telecommunications; U.S.	An investigation of the potential contribution of telecommunications to energy conservation and management

commissions study related topics such as the thermal and visual environmental requirements of building occupants, the external microclimate of buildings, and building energy management systems.

Subgroup 2 of Working Commission W67 on Energy Conservation in the Built Environment is concerned with all aspects of energy use in buildings: its control, its prediction, and its conservation. It has organized several major symposia on this topic, bringing together building researchers, engineering and architectural designers, and other building industry professionals from all over the world. Their Dublin Symposium[16] devoted an entire session to a study of national measures designed to encourage energy conservation in buildings.

The Representatives of the European Heating and Ventilating Associations (REHVA) are responsible for a considerable amount of international inter-professional exchange within their field of activity. Founded in 1963, the objectives of REHVA are ". . . the international promotion of the science and practice of heating, ventilating and air conditioning, and the advancement of education in this field."[17] One of the major vehicles in the pursuit of these objectives is the promotion of an international congress every 4 or 5 years. The 1980 Congress, for example, held in Budapest, took energy conservation in architectural design, and energy-saving principles for the design of heating, ventilating, and air-conditioning systems, as two of its main topics.[18] Although currently covering a nominal 16 European countries, representatives of over 24 nations attended this congress.

The main British and North American professional bodies serving the heating and ventilating industry, the Chartered Institution of Building Services (CIBS), and the American Society of Heating, Refrigerating, and Air-Conditioning Engineers (ASHRAE), held their first joint conference in 1981. Clearly, this type of international exchange is on the increase, with a joint REHVA/ASHRAE/CIB/IIR World Congress on Heating, Ventilation, and Air-Conditioning (the first in this field), taking place in 1985 at Copenhagen.

Other international events of this nature continue to proliferate, as building researchers, academics, engineering and architectural designers, and so on devote increased effort to the better understanding of all aspects of building energy performance.

III. THE ROLE OF NATIONAL INSTITUTIONS

Having looked at the part played by some international agencies at governmental and professional levels we now turn to the role of national institutions. Governmental and professional categories will again be used, and their activities with respect to the drafting of building energy performance standards will be discussed.

A. Governmental Agencies

At the national level, individual governments have been quick to respond to international fuel supply problems. The form of this response has varied according to the particular circumstances of each country. At its most basic, it has involved the setting up of appropriate agencies such as ministries of energy, increased funding of indigenous energy exploration programs, and greater attention being given to an assessment of energy demand. The assessment of energy demand has become a critical issue as far as government policies are concerned, since one of the major tasks facing policy makers is the development of proposals which match that demand to energy supply, and vice versa.

If such proposals are to be properly developed, it is essential to know the energy consumed by each major end-use sector (for example, transport, industry, domestic, and commercial). A reliable breakdown by fuel type would make this information particularly useful. For some of the government funded energy research and development organizations formed in the mid-1970s, the quantification and characterization of the energy demand of their country was a top priority. In most cases it was found that, although its consumption of oil was rather less than that of transport, the building sector (domestic, commercial, and a part of industry) was directly responsible for a significant proportion of the energy demand of each country.

One of the results of this finding was an upsurge of interest in the possible role for building energy performance standards. Of course, other measures such as energy pricing and rationing are possible and have been used successfully in an attempt to regulate energy demand. Unfortunately, energy pricing tends to be applied haphazardly and in a way which discourages long-term planning, while rationing is normally used as a stopgap measure under near emergency conditions.[19] Building energy performance standards, which have been subjected to the scrutiny of a broad spectrum of interested parties, are often seen as a more acceptable means of control than measures which may result from short-term political expediency.

Another outcome of the quantification of the significance of energy demand in the buildings sector was a desire on the part of many governments to (literally) put their own houses in order. Most national governments own or occupy large numbers of buildings and seem to feel obliged to set a good example. Many governments have conservation programs in place and some of these have been dramatically successful. For example, the Property Services Agency of the U.K. virtually halved the energy consumed by government buildings over an 8-year period.[20,21] Clearly, such programs allow government departments to develop a considerable expertise in the measurement of the energy performance of a wide range of buildings, an expertise which in many instances has had a beneficial effect on the drafting of energy performance standards.

B. The Building Professions

Professional bodies with an interest in energy and buildings have played important roles on the national scene. The design professions have been particularly active on behalf of their members. In the U.S. and the U.K., the main building services professional bodies (ASHRAE and CIBS, respectively) have been active in the drafting of building energy performance standards. These have been widely adopted, both in their home and in other countries.

The American Institute of Architects (AIA), through its Research Corporation, has been heavily involved in the work leading to the U.S. Department of Energy draft building energy performance standards. Other professional groups have participated in this work too, and some new groups, such as energy managers or energy engineers, have emerged to play a positive role in this area.

Hand-in-hand with this search for building energy performance standards has been a reappraisal of the methods conventionally used for the prediction and measurement of building energy use. It is felt that if standards related to energy use are to be set and met, then the effects of design decisions should be predictable at the design stage and the results of applying such standards be verifiable. This reappraisal, together with the search for improved and validated techniques and their dissemination and adoption by practicing designers, has involved the education and research establishments as well as the professional organizations (the outcome of their efforts in this direction are explored in more detail in Chapter 6).

Such considerations have awakened the professions to the needs of their members, most of whom received their formal education before energy criteria assumed their present importance in the design process. As a result, a considerable effort is being put into continuing professional development programs. The AIA have embarked upon an ambitious "Energy in Architecture" program aimed at updating the professionally qualified architects of the U.S. in subject areas such as energy analysis and energy-conscious design methods.[22] In the U.K., the Royal Institutes of British Architects (RIBA) and Chartered Surveyors (RICS) together with the Chartered Institution of Building Services (CIBS) are conducting combined educational programs with Department of Energy sponsorship.[23] In New Zealand, a survey has been undertaken of the energy educational requirements of all the building industry professions in an attempt to gear continuing education courses for the perceived needs of the industry as a whole.[24]

The other main sections of the building industry — manufacturers, contractors, property owners, and so on — together with the fuel supply industries, also have an active interest in the energy performance of buildings. Not surprisingly, their motivation can be one of self-interest, or at the very least suspected of being so. Despite this, these sections of the building industry frequently take important initiatives and provide substantial supplementary inputs.

C. Standards Associations

The processes whereby building energy performance standards are formulated varies from country to country. In many cases, they result from a national consensus or cooperative procedure which involves input from a large number of concerned and knowledgeable people, representative of a wide range of interests. This type of approach has been taken by the Standards Association of New Zealand (SANZ). In other cases, several different groups may be pursuing similar goals more or less independently. In the U.S., for example, the federal government, some state governments, and a professional engineers' society have made substantially independent attempts to formulate standards of this type. In other cases still, a particular body may take the initiative, involving the others as appropriate; the CIBS approach would come under this category.

As might be expected, a variety of approaches inevitably results in a variety of types of standards. We shall now examine these in some detail.

IV. BUILDING ENERGY STANDARDS

Having outlined the concerns and activities of the major national institutions in relation to energy and building, we shall now examine the most important manifestation of these concerns, building energy performance standards.

While our main concern will be with the energy performance of the building as a whole, it will be recognized that this performance is dependent on a large number of individual factors. Therefore, before going on to discuss building energy performance standards as a whole, we shall briefly examine those individual factors. They may be classified under four main headings: environmental requirements, building fabric, energy-consuming services, and user activity patterns.

A. The Environmental Requirements of the Occupants

The proper specification of environmental requirements is a key issue in relation to the energy consumed in buildings. Under this heading such matters as thermal comfort, ventilation, and lighting conditions are examined. Standards and other design data on these matters are readily available and generally accepted, but many of them were formulated with specific regard to health and safety, without any special consideration being given to energy conservation. Since the energy performance of a building must ultimately be measured against its environmental performance, a brief appraisal of currently accepted criteria will be presented.

B. The Design of the Building Fabric

This category includes the choice of materials, the shape of the building, its orientation with respect to the sun, its color, thermal mass, insulation, glazing, shading, and so on. All of these can influence energy consumption both at the construction stage and during the life of the building. The practical extent of these influences is described in Chapters 5 and 6, respectively.

With only one exception, none of the factors listed above is the subject of standards. The insulation of the building fabric is that exception, and while there is no doubt that it is a matter of high priority in most building design work, many of the other factors are equally significant. However, the difficulty of setting standards in some of these other areas is considerable, and there is the ever increasing danger of the requirements of different standards coming into conflict with one another.

C. The Energy-Consuming Building Services

The items of equipment which go to make up the heating, ventilating, and air-conditioning systems and the hot water, electrical, and lighting services of buildings, together with their operational and control features, are included in this heading. Since these are the items which consume energy directly, it is obvious that they will have a significant impact, an impact which is discussed in full in Chapter 6.

Suffice it to say here, that although there have been some moves towards standards concerned with the energy performance of such equipment, these have tended to consider individual components rather than the systems as a whole. The possibilities for conflicts between individual components is even more pronounced here than with the elements of the building fabric. Heating, cooling, and lighting systems, for example, can interact closely with each other, with the building fabric, and with external climatic conditions, in ways which defy the drafting and imposition of unsophisticated standards.

Nevertheless, good standards of design must be striven for in respect of the energy performance of both the building fabric and the building services. Even the most energy-conscious operating procedures will be to little avail if the building designers have ignored good energy performance principles.

D. Activity and Occupancy Patterns

This category involves two main areas which impinge on the energy performance of buildings. The first concerns the use of energy-consuming equipment which is directly related to the "productive" activity of the building occupants, as opposed to that which is designed to cater to their environmental requirements — in other words, activity-specific equipment such as electric typewriters, photocopiers, computers, and the like. Scope for energy-related standards exists in this field, but these are outside the scope of this book.

The other main area is that of occupant and operator behavior. While it is clear that poor building design can undermine good operating procedures, it is equally clear that the objectives of an inherently energy-conscious design can be frustrated by poor operating procedures. However, while it may be feasible to advance energy performance standards that deal with matters of building design, one might reasonably speculate that only the foolhardy standard maker would attempt (directly at any rate) to specify occupant behavior.

E. Energy Implications of Building Environment Standards

As alluded to previously, building "energy performance" standards cannot be considered in isolation from building "environment" standards. The energy performance of a building is not satisfactory if the quality of its environment is poor. It is therefore important that the criteria against which environmental quality is measured are securely founded. However, the criteria currently used were formulated some time ago, nearly half a century in some cases. It is therefore important that the basis of these criteria be reexamined in the light of more recent findings and from the point of view of energy conservation. Three main sets of criteria, those relating to thermal comfort, ventilation, and lighting, are reviewed in the following pages.

1. Thermal Comfort Criteria

The four main environmental factors which affect one's sensation of thermal comfort are air temperature, radiant temperature, humidity, and air movement. Many attempts have been made to integrate various combinations of these factors into a single index of thermal comfort. However, the difficulties and inaccuracies involved in taking up to four separate measurements, especially under field conditions, often result in the potential superiority of such indexes remaining unrealized. Air temperature alone is still the most valuable guide to thermal comfort.

To a first approximation, heating energy requirements are directly proportional to the inside-outside temperature difference. The use of an 18°C (64.4°F) inside design temperature, rather than say 20°C (68°F), during a winter with an average outside temperature of 10°C (50°F) would result in a theoretical energy saving of 20%. Supposing half of this was achievable in practice, one could reasonably assume that a significant reduction in heating energy use in the buildings sector would result from such a move. For summer conditions there seems little doubt that savings would similarly accrue if thermal design temperatures were allowed to rise slightly.

These comments merely recognize, and use to advantage, the normal acclimatization processes and seasonal changes in the amount and weight of clothing worn. The important implication from the point of view of building thermal comfort is the realization that it is far too simplistic to specify a single temperature for all seasons, and in addition, that such a practice will result in unnecessary energy use.

2. Ventilation Criteria

Ventilation standards tend to be set in relation to the dilution, to some acceptable level, of any contaminants being released in the occupied space. In this context, a contaminant may range from the products of respiration and occupation (moisture, carbon dioxide, and body odors) to dust, fumes, bacteria, and heat.

For many spaces, the predominant criterion for determining appropriate ventilation rates is the dilution of odors. With only specialized exceptions this results in rates which are far above the level required to avoid biochemical and respiratory hazards. It has been suggested that for general occupancy one might aim for fresh air ventilation rates based on the dilution of carbon dioxide rather than body odors. This would result in rates of around 2 ℓ/sec. ASHRAE Standard 62-1981, ''Ventilation for Acceptable Indoor Air Quality'' suggests various means of reducing the fresh air quantity, but quotes an absolute minimum value of 2.5 ℓ/sec (5 ft^3/min) per person.

There is little doubt that many buildings get considerably more ventilation than is necessary. Much of this is due to ''leaky'' building envelopes and poorly designed and maintained natural and mechanical ventilation systems. Since the energy expended in heating or cooling outside air is proportional (to a first approximation) to the product of the ventilation rate and the inside-outside temperature difference, it is clear that ventilation criteria are of great importance in relation to the energy performance of a building.

3. Lighting Criteria

The traditional recommended lighting standards for workplaces are based upon two notions. First, that the things that people need to be able to see to do their work are arranged upon a horizontal plane. Second, that the illuminance of this plane can be quite simply related to the difficulty of the visual content of the work to indicate the adequacy of the lighting. Both these notions have been subjected to critical examination and reappraisal during recent years.

This reappraisal has provided the opportunity, not only for the contribution of daylight to the general lighting of buildings to be more realistically assessed, but also for more valid comparisons of the efficiencies of alternative forms of electric lighting to be made. The fashionable ''downlight'' may satisfy standards of horizontal illuminance, but it is a comparatively low efficiency form of lighting.

With regard to the relationship between illuminance and the visibility of tasks, several factors other than illuminance can affect task visibility. The most important of these is the extent to which the lighting reveals the contrast by which the task detail is seen.

These new ideas have resulted in alternative ways of specifying lighting standards, which offer the potential for substantial energy savings. A recent review[26] describes several cases where this has been achieved.

F. Classification of Building Energy Performance Standards

Before attempting to review the different approaches that have been taken to the formulation of building energy performance standards, a broad system of classification will be presented. This will enable readers to perceive the conceptual bases of various approaches and judge for themselves the relevance of any new standards. It must be emphasized that the approaches and standards reviewed here represent a snapshot of what is a rapidly developing situation. New standards are being evolved at an increasing rate, previous approaches are being discarded or modified in the light of experience, and recently formulated standards are already being updated.

The types of building energy performance standard which are currently advocated may be broadly classified into "prescriptive" and "performance" categories. In general terms, the dividing line between a prescriptive and a performance standard is very much determined by the definition of the system boundaries. A prescriptive standard will normally apply to the individual components of a system; a performance standard to the system as a whole.

It should of course be appreciated that prescriptive standards are performance standards for individual components. For example, the specification of the insulation value for a wall may be regarded as a performance standard as far as the wall itself is concerned. However, in terms of the building of which the wall is part, such a value would be regarded as a prescriptive standard.

Standards in the performance category may usefully be subdivided into those appropriate to the design or "prediction" stage and those suitable for the in-use or "monitored" stage of the life of a building.

1. Prescriptive Standards

In the context of the building as a whole, the term prescriptive would be applied to any standard that related to one or other of the various components of the building.

The Provisional New Zealand Standard (NZS 4218P : 1977) "Minimum Thermal Insulation Requirements for Residential Buildings"[27] is an example of a prescriptive standard in the context of whole building energy conservation. It specifies the thermal insulating properties required of the building fabric, allowing some trade-off between different parts. This particular standard also embodies a performance option, the first of its kind for residential buildings.

In North America, the prescriptive standard which has gained the greatest acceptance is the "ASHRAE Standard 90-75 Energy Conservation in New Building Design."[28] This document was initiated at the National Bureau of Standards as an interim standard until such time as a full performance approach would become feasible. While essentially a prescriptive-type standard, its uniqueness lies in the fact that it attempts to specify the energy requirements of all the main building components in a single document, with chapters on:

- Exterior envelope
- HVAC systems
- HVAC equipment
- Service water heating
- Electrical distribution systems
- Lighting power budget determination procedure

The final sections of the standard are entitled: "Energy Requirements for Building Designs Based on Systems Analysis" and "Requirements for Buildings Utilising Solar, Wind or Non-Depleting Energy Sources". These indicate that one is allowed to develop a nonstandard design, providing the predicted annual energy consumption is no greater than that of a standard design and that renewable energy sources need not be included in the analysis. In other words, progress has been made towards what might ultimately become a completely performance oriented approach.

The essential features of ASHRAE 90-75 have now been embodied in the new "Code for Energy Conservation in New Building Construction" promulgated by the U.S. National Conference of States on Building Codes and Standards (NCSBCS).[29] This document specifies design conditions as well as energy conservation guidelines and its format is restructured with respect to ASHRAE 90-75, putting the "performance" approach first, followed by the "prescriptive"; an "acceptable practice" approach is also allowed for some types of building.

The content of ASHRAE 90-75 was also embodied into a document issued by the Canadian Associate Committee on the National Building Code (ACNBC). This was entitled "Measures for Energy Conservation in New Buildings — 1978,"[30] the word "measures" serving to indicate that the Canadians did not intend immediate adoption of the provisions of the document as a code. The Standing Committee on Energy Conservation in Buildings of the Canadian ACNBC has set up task groups to investigate energy consumption in different categories of buildings.[31] Clearly, they anticipate moving towards performance-type standards.

Hence, it would appear that although many code makers initially settled for prescriptive, if fairly comprehensive, energy conservation measures, their ultimate aim was for a performance-type standard. The reasons for this are not difficult to see. Prescriptive-type standards tend to embody the concepts of current practice and conventions. While this can provide a useful check for the unadventurous, it can also stifle innovative solutions. Even fairly common building types, such as air-conditioned office blocks, defy the simple prescriptive formula. The whole-building performance approach was seen early on as the ultimate goal, allowing freedom for innovation and scope for trade-offs between the different components of the building system.

2. Performance Standards

This type of standard is concerned with the energy performance of the building as a whole, as opposed to the individual parts. Many types of performance standard have been considered,[32] but the prediction type and the monitored type, appear to have gained most acceptance.

a. Prediction Performance Standards for New Buildings

These have two principal requirements. One is the specification of acceptable energy consumption design "targets" appropriate to particular circumstances (e.g., activity type to be housed, hours of occupancy, climatic location of the building, etc.). The other is the precise specification of the method to be used to predict energy use, for comparison with the targets. These standards are not directly concerned with the actual energy consumed by the building.

b. Monitored Performance Standards for Existing Buildings

As with "prediction" performance standards, this type also requires the specification of energy consumption targets. In this case, however, the actual consumption is monitored by specified methods, and compared with the target. This is a direct method which can be applied to whole buildings and includes "real world" operation and control factors.

c. Energy Consumption Targets

Both the prediction and the monitored performance standards require energy consumption target figures to be set. Prediction performance targets may be set in terms of annual energy consumption per unit floor area (the AEUI) and related to the activity housed, the climate, hours of operation, and so on. For heated, naturally ventilated buildings, a "design energy target" specified in terms of the design heat loss per unit floor area has been put forward.[33]

A monitored performance standard, on the other hand, would be concerned with the annual energy consumption of the real building under actual operating conditions. An AEUI target could be set and allowance made for activity housed, climatic zone, and so on.

Several methods of setting both types of energy consumption target have been suggested and implemented over the last few years. These range from the setting of percentage reductions to be achieved for existing buildings to the specification of maximum allowable consumption figures. Such targets can be applied at the design stage, or when the building is in operation.

The British Property Services Agency (PSA), for example, took some account of relative building size and major climatic differences in setting provisional targets for naturally ventilated offices.[34] The building size is expressed as the ratio of the floor area to the external fabric area. The following are examples of targets suggested for heating and hot water service only:

	Energy consumption at building boundary		
	$MJ\ m^{-2}$ $year^{-1}$	$Btu\ ft^{-2}$ $year^{-1}$	Floor/fabric ratio of the building
New offices	530	46,640	1.0
	411	36,170	1.5
	355	31,240	2.0
Existing offices	1274	112,100	1.0
(with 40% glazing)	1097	96,540	1.5
	1020	89,760	2.0

These figures apply for the southern part of Britain. They are multiplied by a factor of 1.13 for the Midlands area and 1.24 for Scotland, as a broad allowance for climate. Electricity for light and power is specified at $119\ MJ\ m^{-2}\ year^{-1}$ ($10,470\ Btu\ ft^{-2}\ year^{-1}$), no matter the floor/fabric ratio.

Owens[35] has attempted to take account of intermittent use. For example, he suggests that a 1000-m^2 ($10,760\ ft^2$) commercial building should have a total energy budget of around $350\ MJ\ m^{-2}\ year^{-1}$ ($31,000\ Btu\ ft^{-2}\ year^{-1}$) with 50% intermittent use; $600\ MJ\ m^{-2}\ year^{-1}$ ($52,800\ Btu\ ft^{-2}\ year^{-1}$) if used continuously.

3. Summing Up

It is appropriate at this stage to attempt an assessment of the impact of building energy standards on the consumption of energy in buildings. The effectiveness of prescriptive standards, which deal only with particular elements of the building fabric (typically, its thermal insulation properties) or the energy-consuming systems (ratings of appliances for instance) is not simple to assess. It seems evident that such standards are insufficient in themselves to ensure a significant reduction in energy consumption. Survey data available from the domestic sector suggests that around 50% of the potential savings due to better insulation may be achievable in practice, but even this evidence is equivocal.[36] Indeed, given the lack of reliable construction data on the existing building stock, it is difficult even to see how one might predict energy savings in a convincing manner. It is interesting to note that only a few authors at the 1982 CIB-W67 Symposium were prepared to estimate the potential savings from the various prescriptive measures described.

It must be admitted that there is, as yet, little evidence that performance standards would fare any better. However, they do possess two overwhelming advantages, in theory at any rate. First, the predicted energy use of a new design can be checked, using a well-documented and unambiguous method of calculation, and compared with a design target. Second, in the case of an existing building, the actual energy consumption can be checked directly, and again compared with an appropriate target.

Provided that the method of predicting energy use has been validated, that targets have been set at realistic levels, and that the current distribution of energy consumption in the buildings sector is known, then it should be a relatively straightforward task to quantify, with some precision, the potential savings that would accrue from building energy performance standards. Of course, life, particularly in buildings, is never as simple as that and neither are energy standards.

Those involved in the drafting of standards have taken a variety of approaches to the formulation of building energy performance standards. Some of these approaches will be described in the next section.

G. National Approaches to Energy Performance Standards

The general aim of the performance approach is to encourage the practice of specifying materials, components, and systems in terms of what they are required to do, rather than what they should be. The particular objective of an energy performance standard is to specify what is required of the building, in terms of a target energy consumption.

Although the principle appears to be a sound one and has won wide acclaim, putting it into practice is not quite as simple as it might at first appear, especially when attempting to consider the building as a whole. Until there are adequate means, both for predicting performance and for checking that the required standard has been achieved and maintained, performance standards will be of little use. At this stage in their development it is frequently found that our current knowledge is not adequate and that many of our prediction techniques are not sufficiently accurate. This is certainly the case as far as building energy performance standards are concerned. Much of the early work in this direction has been devoted to the filling of knowledge gaps concerning the energy used by existing buildings and the development and evaluation of methods for predicting energy use.

In the following sections the approach taken by a number of organizations to problems related to the drafting of energy performance standards will be described. The organizations selected were

1. The American Society of Heating, Refrigerating and Air-Conditioning Engineers
2. The U.S. Department of Energy
3. The California Energy Commission
4. The Chartered Institution of Building Services (U.K.)
5. The Standards Association of New Zealand

The basis of this selection has been to allow a broad spectrum to be presented, based on size of country, type of organization, current progress, and uniqueness of approach. A detailed description of content will not be given. All the standards involved are available in print, some of them have substantial back-up materials in the form of published reports, and most have been the subject of extensive discussion and comment in the relevant literature. Rather, the broad approach and general content will be outlined, the principles involved reported, and the main features of each highlighted.

1. The American Society of Heating, Refrigerating and Air-Conditioning Engineers

As a society, ASHRAE has produced many standards related to the needs of their members and the industry they serve. Some of these standards have been approved by the American

National Standards Institute and have been widely adopted by different authorities in the U.S. and elsewhere. Since their members are involved in the design and operation of much of the energy-consuming equipment installed in buildings it is natural that ASHRAE should be in the forefront of the drafting of standards concerned with energy use in buildings.

Its pioneering "ASHRAE Standard 90-75: Energy Conservation in New Building Design" has been referred to earlier.[28] The standard has now been revised,[37] but remains an essentially prescriptive document as far as the building as a whole is concerned.

Despite the overall prescriptive character of this standard, one of its sections (No. 10) does allow for nonconforming building designs to be assessed on a performance basis. Briefly, if the design does not conform to the prescriptive standards laid down, then its annual energy consumption must be calculated and shown to be less than or equal to that of a conforming building. In a sense, the figure for the annual energy consumption of the conforming building has become the performance target.

A further section embodies the concept of Resource Utilization Factors (RUFs). These factors are designed to account for the energy consumed in processing, refining, and delivering each fuel from point of extraction to the building boundary. The question of the inclusion of such factors is one which has exercised many groups. ASHRAE has treated the matter in some depth, computing sample values for each type of fuel for different regions of the U.S. and for different seasons of the year, as appropriate. However, at the time of writing it would appear that this section is likely to be dropped.

Standard 90-75 (or 90A-1980, as revised) deals with new building design, but the same general approach to energy targets is also evident in ASHRAE attempts to draft energy standards for existing buildings. A series of such drafts (the 100 Series) has emerged dealing with low rise residential,[38] high rise residential,[39] commercial,[40] industrial,[41] institutional,[42] and public assembly[43] buildings. These standards are at various stages of development.

Their main objective appears to be to encourage energy conservation by making the carrying out of energy audits a necessary prerequisite for compliance with the standard. The target is specified as the annual energy consumption of a similar new building complying with the requirements of ASHRAE 90-75. Thus each existing building will have its own individual energy consumption target. At the time of writing, these drafts are under review, and it is difficult to predict the form in which they may gain final approval.

It will have become evident that the precise measurement and expression of building energy performance is of paramount importance. ASHRAE has tackled this problem too and has drafted an appropriate standard. Designated ASHRAE Standard 105P,[44] this document deals with the measurement of energy use in existing buildings and the expression of the energy consumption of a building.

To sum up the pioneering efforts of ASHRAE in this field: a comprehensive prescriptive standard for new buildings has been produced, attempts have been made to take account of energy resource utilization factors, and the need for a performance option has been recognized. The requirements of existing buildings have also been recognized, and standards which utilize prediction and monitored performance elements, combined with prescriptive requirements (a curious and possibly confusing combination of principles) have been produced. Standard methods for measuring and expressing building energy performance have also been proposed. Overall, it is a remarkable effort, based almost entirely on the voluntary efforts of ASHRAE members. Their efforts have been emulated and their standards adapted for use throughout the U.S. and elsewhere.

2. The U.S. Department of Energy

One of the responses of the U.S. Federal Government to the energy crisis was the setting up of a Department of Energy. Another, of particular concern to the building community, was the passing of an act[45] concerned with the energy performance of new buildings. The act called for the development of what have become known as the Building Energy Performance Standards, commonly abbreviated to BEPS.

As proposed in 1979 this standard is of the prediction performance type. In other words, it is concerned with the performance of the building as a whole, as predicted at the design stage. The requirement of the proposed standard is simply that "The design energy consumption of the building design of a new building shall not exceed its design energy budget." Methods for determining both the budget and the consumption figures are specified.[46]

The text of the proposed standard is fairly brief, comprising about four pages of text mostly taken up with the definition of technical terms and the classification of building types, together with four tables specifying energy budget levels for residential and commercial buildings of various kinds in a range of climatic zones and weighted according to the relative value, to the nation, of different fuels. Table 3 gives some typical values.

Despite its apparent brevity, a considerable amount of documented effort lies behind this proposal. Apart from the 50 or so pages of explanatory material contained in the *Federal Register*,[47] ten technical support documents,[48-57] which together run to well over 2000 pages, have been published. These documents represent the culmination of the considerable research program of work undertaken by a large number of groups on behalf of the Department of Energy. The foundation of this program was laid with a survey of a large number (1661 returns) of post-1973 building designs, followed by the redesign of 168 of these using energy conservation principles. The results of energy analyses of this sample of buildings provided the basic source for the design energy budget figures. All in all a massive and costly program, especially by comparison with the largely voluntary exercise of ASHRAE.

As one of the first attempts at a performance-type standard, it was inevitable that the BEPS program would attract a great deal of comment at the various stages in its development. Very few standards have received quite so much attention in the architectural and the engineering press.[58,59] While it would appear that few would quarrel with the principle underlying this type of standard, a considerable amount of criticism has been leveled both at the bases of the budget figures and at the specification of computer methods for calculating the design energy consumption of a proposed building. As might be expected, the various energy suppliers are particularly concerned with the likely impact of the fuel weighting factors on their respective industries. ASHRAE concerns relate, inter alia, to the method of implementation of such a standard and the potential complexity of the computational procedures involved. The American Institute of Architects on the other hand, though fully aware of all these problems (which they see as soluble) is strongly committed to the performance approach.

To summarize, the proposed standard is of the prediction performance type, applicable to new buildings. The energy budget figures are based on the designed energy performance of real post-1973 buildings and take account of building activity, climatic zone, and the national value of the fuels used. Computer-based methods and standard operating conditions are specified for calculating the design energy consumption. Despite the considerable effort that has gone into the formulation of this standard, it appears unlikely that it will pass into general use.

3. The California Energy Commission

Having seen the approach of the U.S. Federal Government to these matters, it is of interest to study how the state of California has tackled the issue of energy standards for buildings. In some respects, the Conservation Division of the California Energy Commission appears to have made considerable progress towards an enlightened blend of prescriptive and performance requirements.

Two standards have been published, one relating to new residential buildings,[60] the other concerned with new nonresidential buildings.[61] The former document is wholly prescriptive

Table 3
ENERGY BUDGET LEVELS[a] FOR COMMERCIAL AND MULTIFAMILY RESIDENTIAL BUILDING DESIGNS (U.S. DEPARTMENT OF ENERGY BEPS VALUES IN MBtu ft^{-2} YEAR^{-1} OR Btu 10^3 ft^{-2} YEAR^{-1d}

State	SMSA[c]	Clinic	Community center	Gymnasium	Hospital	Hotel, motel	Multifamily high rise	Multifamily low rise	Nursing home	Office large	Office small	School elementary	School secondary	Shopping center	Store	Theatre/auditorium	Warehouse[b]
Alabama	Birmingham	123	107	127	353	166	114	110	161	113	101	89	117	181	142	139	53
	Mobile	142	129	147	406	192	127	132	187	131	116	96	133	207	166	162	47
Arizona	Phoenix	146	133	152	406	196	131	136	192	134	119	100	137	212	171	168	49
California	Bakersfield	123	109	127	358	167	113	112	162	113	100	86	116	181	143	140	48
	Fresno	120	105	123	353	163	112	108	158	111	98	85	114	178	139	136	50
	Los Angeles	112	101	115	364	157	103	103	151	106	91	74	106	171	132	126	42
	Oakland	108	93	108	353	150	102	94	143	101	87	75	103	164	125	119	50
	Sacramento	118	102	120	353	160	110	104	154	108	96	84	112	175	136	132	52
	San Diego	114	103	117	364	158	104	106	153	107	92	75	107	172	134	128	40
	San Francisco	108	92	109	353	150	103	94	143	101	87	76	103	165	125	119	51
Colorado	Denver	122	98	123	338	162	119	100	156	109	100	97	118	178	137	135	71
Connecticut	Bridgeport	128	105	130	353	170	123	106	156	115	105	100	123	186	144	142	71
	Hartford	125	101	127	338	165	122	102	159	112	103	100	121	181	140	139	74
D.C.	Washington	127	107	129	353	169	120	109	164	115	104	96	121	185	144	142	63
Florida	Jacksonville	143	130	149	406	193	128	134	189	132	117	97	134	209	167	164	47
	Miami	152	142	161	406	203	133	147	201	140	125	103	141	219	179	178	41
	Tampa	145	135	152	406	196	129	139	193	135	119	98	136	212	171	168	43
Georgia	Atlanta	122	106	125	353	165	114	108	160	112	100	88	116	180	141	138	53
Idaho	Boise City	124	100	125	338	163	120	101	158	111	101	98	120	179	139	137	71
Illinois	Chicago	127	102	129	338	167	124	103	161	113	104	103	123	183	142	141	75
	Glenview	129	103	130	338	168	125	105	163	114	105	103	124	184	143	143	75
Indiana	Indianapolis	128	103	130	353	168	124	105	162	114	105	102	123	184	143	142	73
Kansas	Dodge City	133	109	135	353	175	128	111	162	119	109	105	128	191	150	149	72
Kentucky	Louisville	128	107	131	353	170	122	109	165	116	105	98	123	186	145	143	66

Louisiana	Baton Rouge	142	129	147	406	192	128	132	188	131	116	97	133	208	166	163	48
	Lake Charles	144	130	149	406	194	130	133	189	133	118	100	135	210	168	165	51
	New Orleans	144	129	149	406	194	130	133	189	132	118	100	135	210	168	164	52
Maine	Portland	130	100	131	335	169	129	101	162	114	107	109	127	186	143	143	86
Massachusetts	Boston	125	101	126	338	165	121	102	159	111	102	99	121	181	140	139	72
Michigan	Detroit	129	103	130	338	168	126	104	163	114	106	105	125	185	143	143	77
Minnesota	Minneapolis	142	109	144	335	180	140	110	175	123	117	122	138	198	155	157	93
Mississippi	Jackson	127	113	131	358	171	117	115	167	117	104	90	120	186	147	145	50
Missouri	Columbia	132	109	134	353	174	126	111	161	118	108	103	127	190	149	148	71
	Kansas City	133	110	136	353	175	127	112	162	119	109	104	128	191	150	149	70
	St. Louis	133	110	136	353	175	128	112	163	119	109	105	128	192	150	149	72
Montana	Great Falls	131	102	132	335	170	129	102	163	115	107	110	127	186	144	144	85
Nebraska	Omaha	130	105	132	335	170	126	106	164	115	107	105	126	186	145	145	76
Nevada	Las Vegas	130	115	135	358	174	118	118	170	119	106	92	122	188	150	148	49
New Jersey	Newark	129	107	131	353	171	123	108	165	116	105	99	124	187	146	144	68
New Mexico	Albuquerque	127	107	129	353	169	121	108	164	115	104	96	122	185	144	142	64
New York	Albany	131	102	132	335	170	129	103	164	115	108	109	127	187	145	145	83
	Binghamton	133	103	135	335	172	132	104	166	117	110	113	130	189	147	147	88
	Buffalo	129	101	130	338	168	127	102	162	114	106	106	125	185	143	142	80
	New York	126	105	128	353	168	120	107	162	114	103	96	121	184	143	141	66
N. Carolina	Raleigh	124	106	127	353	167	117	108	161	113	101	92	119	182	142	139	59
N. Dakota	Bismarck	146	110	147	335	184	146	111	179	125	121	129	143	203	158	161	102
Ohio	Akron	128	102	129	338	167	125	103	161	113	105	104	124	183	142	141	77
	Cincinnati	130	107	132	353	172	124	109	166	117	106	101	125	188	147	145	70
	Cleveland	129	103	131	338	169	126	104	163	114	106	105	125	185	144	143	78
	Columbus	128	103	130	338	168	125	104	162	114	105	103	124	184	143	142	75
Oklahoma	Oklahoma City	129	110	132	353	172	121	112	167	117	106	97	123	187	147	146	61
	Tulsa	127	109	130	353	170	119	111	165	116	104	95	121	185	146	144	59
Oregon	Medford	120	99	121	353	162	116	101	155	109	98	91	116	177	136	133	64
	Portland	119	98	120	353	161	116	99	154	108	97	91	115	176	135	131	66
Pennsylvania	Allentown	129	105	131	353	171	125	106	158	116	106	102	125	187	145	144	74
	Philadelphia	131	107	133	353	173	126	109	160	117	107	102	126	189	147	146	71
	Pittsburgh	126	101	127	338	165	122	103	159	112	103	100	121	181	141	139	72
S. Carolina	Charleston	124	110	128	358	168	114	113	163	114	102	88	118	183	144	141	49
Tennessee	Memphis	126	109	129	353	169	117	111	164	115	103	92	120	184	145	142	56
	Nashville	125	107	128	353	168	117	109	162	114	102	92	119	183	141	141	58

Table 3 (continued)
ENERGY BUDGET LEVELS[a] FOR COMMERCIAL AND MULTIFAMILY RESIDENTIAL BUILDING DESIGNS (U.S. DEPARTMENT OF ENERGY BEPS VALUES IN MBtu ft^{-2} YEAR^{-1} OR Btu 10^3 ft^{-2} YEAR^{-1})[d]

State	SMSA[c]	Clinic	Community center	Gymnasium	Hospital	Hotel, motel	Multifamily high rise	Multifamily low rise	Nursing home	Office large	Office small	School elementary	School secondary	Shopping center	Store	Theater/auditorium	Warehouse[b]
Texas	Amarillo	126	106	129	353	168	120	108	163	114	103	95	121	184	144	141	63
	Brownsville	150	139	157	406	200	132	143	198	138	123	101	139	216	176	174	43
	Dallas	131	116	136	358	175	119	119	171	120	107	94	124	190	152	150	50
	El Paso	126	110	129	358	169	116	119	164	115	103	90	119	186	145	142	52
	Fort Worth	128	113	132	358	171	117	116	167	117	104	90	120	186	148	145	50
	Houston	145	130	150	406	195	130	134	190	133	118	100	136	211	169	166	51
	Lubbock	126	107	128	353	168	118	110	163	114	103	93	120	183	144	141	58
	San Antonio	146	131	151	406	196	132	134	191	134	119	102	137	212	170	167	53
Utah	Salt Lake City	129	104	131	338	169	125	105	163	114	106	104	125	185	144	143	76
Vermont	Burlington	134	103	135	335	173	133	104	167	117	110	114	131	190	147	148	89
Virginia	Norfolk	123	105	125	353	165	115	108	160	112	100	90	117	180	141	138	56
	Richmond	129	107	131	353	171	122	109	165	116	105	98	123	186	146	144	66
Washington	Seattle	119	96	119	353	160	116	97	153	107	96	91	115	176	134	130	69
	Spokane	126	99	126	338	165	124	100	158	111	103	103	122	181	139	138	79
West Virginia	Charleston	128	106	130	353	170	123	108	164	115	105	99	123	186	145	143	68
Wisconsin	Madison	131	102	130	335	170	130	103	164	115	108	110	128	187	145	145	84
	Milwaukee	131	102	132	335	170	129	103	164	115	108	110	128	187	145	145	84
Wyoming	Cheyenne	128	100	129	338	167	127	101	161	113	105	106	125	184	142	141	82

a Figures include design energy requirements for heating, cooling, domestic hot water, fans, exhaust fans, heating and cooling auxiliaries, elevators, escalators, and lighting.

b 1 MBtu ft^{-2} year^{-1} = 11.4 MJrm^{-2} year^{-1}.

c SMSA is the Standard Metropolitan Statistical Area.

d Space is reserved in this table for restaurants and industrial buildings.

in nature, and one can readily appreciate the reasoning for this type of approach in the case of residential buildings. Nonresidential buildings, on the other hand, are expected to meet a prediction performance standard, the corresponding regulations representing the first attempt to establish energy standards using computer analysis. Considerable resources were expended in their creation, as is evident by the many reports recounting the details of their development and application that have been published.[62-67]

As with the proposed BEPS of the Department of Energy, design energy budget figures are specified with respect to a range of building activities and climatic zones. Provided the design and construction of a building comply with a small number of specific regulations (mostly related to air leakage, controls, pipe and coil insulation, hot water service, and electricity distribution) then the only other requirement is that its " . . . total calculated annual energy consumption . . . does not exceed the product of the gross floor area of conditioned space . . . and the allowable energy budget . . . "

As part of the lead-up to these design energy budgets, expert panels agreed on "model designs" for nine building types: large and small offices, large and small retail stores, supermarkets, restaurants, public assembly buildings, schools, and warehouses. Each building was specified both on the basis of conventional 1974 practice and on the basis of the ASHRAE 90-75 standard (which the draft Californian regulations embodied). Sixteen climatic zones (see Figure 1) were delineated for the state of California, some of these being amalgamated to form five zones for regulation purposes. Three types of conditioning systems were scrutinized, viz., heating only, cooling only, or both heating and cooling. Occupancy and other operational parameters were specified.

The CAL/ERDA computer program was then used to calculate the design energy budget figure for each of the nine building types, under both design standards (i.e., 1974 conventional and ASHRAE 90-75) in each of the climatic regions. The energy budget figures derived from these analyses are reproduced on Table 4. It should be noted that the electricity consumption is multiplied by a factor of three to allow for generating efficiency.

These figures relate to energy used for comfort heating, comfort cooling, ventilation, hot water service, lighting and a 0.5 W ft^{-2} (approximately 5 W m^{-2}) allowance for miscellaneous equipment use. The influence of process loads is not considered, although their effect is recognized. It is also worth noting that the capital costs of all designs were professionally estimated and the present value life-cycle costs calculated. Complying with the standard was found to save energy costs in every case studied.

As well as being used to develop the budget figures, the CAL/ERDA computer program has been specified for use in checking compliance with the standard. Other programs have received approval too[68] and where buildings with occupancies of less than 300 persons are involved, a simplified manual procedure will be permitted.

Once again, we have a standard of the prediction performance type, intended for use with new nonresidential buildings. This time the energy budget figures are based on the computed energy needs of model buildings designed to meet the requirements of ASHRAE 90-75. The energy budget figures are specified with respect to 5 climatic zones, 14 building activities and loadings, and whether the space involved is heated, or cooled, or both; and a weighting factor of three has been applied to the case of electricity use. The designer may select from a range of approved computer programs, or employ a manual method if the building occupancy is less than 300.

4. The Chartered Institution of Building Services (U.K.)

As with ASHRAE in the U.S., the Chartered Institution of Building Services of the U.K. has been very concerned with different aspects of energy conservation in buildings. Part of this concern has manifested itself in the production of the CIBS Building Energy Code. This unique document has been published in four parts:[69]

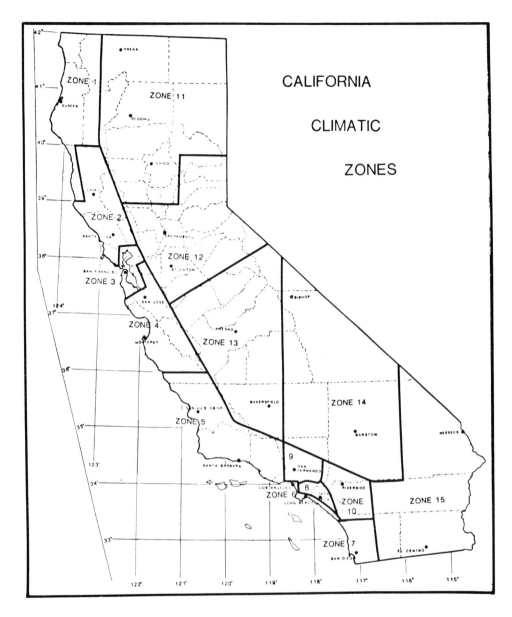

FIGURE 1.   California climatic zones. (Reproduced by permission of the California Energy Commission).

Part 1:   Guidance towards energy-conserving design of buildings and services
Part 2:   Calculation of energy demands and targets for the design of new buildings and services
Part 3:   Guidance towards energy-conserving operation of buildings and services
Part 4:   Measurement of energy consumption with targets for existing buildings and services

Parts 1 and 3 of the code present energy conservative design and operational guidance, respectively. Part 1 is fairly prescriptive in character while Part 3 has a checklist-type format. Parts 2 and 4 are performance oriented, the former being of the prediction type, the latter the first monitored performance type to be considered here.

Part 1 gives what might be termed prescriptive guidance in respect of new building design,

Table 4
TABLE OF CALIFORNIAN NONRESIDENTIAL BUILDING ENERGY BUDGET FIGURES

(Thousands of British Thermal Units Per Gross Square Foot of Conditioned Floor Area)

Ref. No.	Occupancy UBC occupancy[a]	Occupant load	Climatic thermal[b] zones 1—5 Heated and cooled	Heated only	Cooled only	Climatic thermal zones 6, 8, 9, 10 Heated and cooled	Heated only	Cooled only	Climatic thermal zone 7 Heated and cooled	Heated only	Cooled only	Climatic thermal zones 11, 12, 13 Heated and cooled	Heated only	Cooled only	Climatic thermal zones 14, 15 Heated and cooled	Heated only	Cooled only
1	(A) Drinking and dining establishments		131	126	82	108	102	82	104	103	80	132	119	96	148	109	126
2	All others		159	154	64	123	114	68	118	106	71	155	140	86	189	120	141
3	B-1		180	171	163	191	163	184	189	162	184	196	173	185	243	168	236
4	Offices	Under 300	141	135	134	145	133	142	145	132	143	148	139	140	165	137	160
5		Over 299	126	125	124	129	128	128	131	130	131	134	131	130	132	129	128
6	Retail grocery store		214	212	167	194	189	176	192	187	176	236	215	199	255	204	229
7	(B-2) Other retail stores	Under 300	180	171	163	191	163	184	189	162	184	196	173	185	243	168	236
8		Over 299	200	195	190	207	195	202	209	196	204	213	199	205	231	205	225
9	Drinking and dining establishments		131	126	82	108	102	82	104	103	80	132	119	96	148	109	126
10	Classrooms		120	118	77	105	94	82	101	94	83	156	143	89	142	97	123
11	Storage		104	104	104	65	65	65	63	63	63	92	92	92	80	80	80
12	B-3, B-4, H		104	104	104	65	65	65	63	63	63	92	92	92	80	80	80
13	E		120	118	77	105	94	82	101	94	83	156	143	89	142	97	123
14	J[d]		—	—	—	—	—	—	—	—	—	—	—	—	—	—	—
15	Others[e]		141	135	134	145	133	142	145	132	143	148	139	140	165	137	160

Note: Convert kilowatt hours (KWh) to British thermal units (Btu) using a conversion factor of 10,239 Btu per KWh. 1000 Btu ft^{-2} year^{-1} = 11.4 MJ m^{-2} year^{-1}

[a] Occupancy, as defined in the Uniform Building Code, 1976 Edition. Occupancies A and B-2 have been further subdivided. For B-2 type occupancies not listed, use the subdivision which most closely describes the occupancy.

[b] Climatic thermal zone, as described in Figure 1.

[c] Use appropriate columns for buildings that are heated and cooled, heated only, or cooled only.

[d] Maximum allowable energy consumption values are under development for UBC occupancy.

[e] Occupancies which are not included in any of the listed UBC occupancies.

From Energy Conservation Standards for Nonresidential Buildings, Conservation Division, California Energy Commission, July 1982.

and acknowledgment is given of the strong influence of the ASHRAE 90-75 Standard. As in the American document, each energy-related component of the total building system, from site factors and building form to ventilation and artificial lighting, is considered in turn. In this case of course, the measures suggested relate to U.K. conditions and design practices.

Part 3 deals first of all with the general planning and implementation of an energy conservation program for existing buildings, together with the requirement for energy surveys and audits. It then describes, in checklist fashion, conservation measures related to the building, its systems and equipment, and their control and operation. Sections on heat recovery and alternative energy sources are included, as well as a summary of investment appraisal procedures.

In their attempt to produce a prediction performance standard, the authors of Part 2 of the CIBS Building Energy Code have come up with what they term a "quite novel" and "therefore untried" method. In common with all prediction-type standards, design energy targets must be specified (e.g., the "design energy budgets" of the U.S. Department of Energy BEPS) and a standardized method for calculating the design energy demand (e.g., BEPS design energy consumption) must be agreed.

As far as the method of calculation is concerned, CIBS has decided to adopt a "conventional" technique (i.e., not computerized). The technique is a manual one which involves the summation of the space heating and hot water service energy use (defined as the thermal demand) together with the lighting and equipment energy consumption (defined as the electrical demand) for standardized occupancy patterns in a range of building types. The use of natural daylighting is taken into account. The technique itself is applicable to heated, naturally ventilated buildings at this stage. A further section (Part 2b) will deal with mechanically ventilated and air-conditioned buildings. As in the American standards, primary energy fuel factors are used. These range from just over unity for coal, oil, and natural gas to a figure of 3.82 for electricity, reflecting current U.K. energy supply conditions.

The demand targets have been established on the basis of this standardized calculation technique. The thermal demand target values for a range of building types are given in Figure 2. These are presented as a function of the so-called Building Envelope Number, a dimensionless parameter which depends on the size, structure, and thermal properties of the building. The electrical demand target is the sum of the equipment power (taken as 10% of the thermal demand target) and the lighting power requirements; figures for the latter are given in Table 5. The sum of the thermal and electrical components represents the total demand target. Compliance with the code simply requires that the calculated demand does not exceed the target, together with the proviso that " . . . the building and its services do not produce extremes of internal temperature in summer, or winter."

The main objectives of Part 4 of the code are to provide performance indicators from which the energy consumption target of an existing building may be assessed and to give guidance on energy measurement and performance monitoring. The performance indicators for a range of building types are given in Table 6. The indicators are based on the results of a survey of 623 buildings.[70] The data has been rounded off to some extent but generally speaking a "good" performance indicator implies a figure in the lower quartile of the sample, while a "poor" or "very poor" performance indicator implies energy consumption in the upper quartile. It was felt that such figures, based on the current consumption of similar buildings, would provide realistic and acceptable targets for existing buildings. Table 7 lists the correction factors which may be applied for nonstandard conditions. Of course, the resulting energy consumption targets are very much related to U.K. climatic conditions and building practices.

This part of the code also details appropriate methods for measuring the energy consumption of existing buildings so that actual performance may be checked against the target

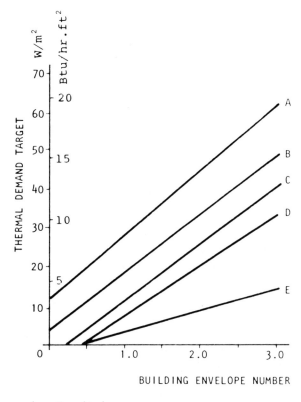

A Hospitals
B Residential buildings
C Hotels
D Offices, shops (5 day)
E Factory (5 day, single shift)

FIGURE 2.   Thermal demand targets for various building types.
(From CIBS Building Energy Code, Part 2(a), Chartered Institution
of Building Services, London, 1981.)

figure. In addition, methods of continuously monitoring the energy performace of existing buildings, involving the use of the thermal performance line, are described (these will be discussed in more detail in Chapter 7).

The following recapitulates the building energy performance aspects of the CIBS approach. First of all, they have assembled energy conservation guidance for the designers of new buildings and the operators of existing ones. Second, they have produced a prediction performance standard for new building designs based on a manual calculation technique and applicable to heated, naturally ventilated buildings: standard operating conditions and fuel factors are specified. Third, a monitored performance standard has been produced with target figures related to the lower quartile energy consumption of a sample of similar buildings. Overall, it is a very comprehensive published package for those involved in building energy conservation.

5. The Standards Association of New Zealand

The Standards Association of New Zealand (SANZ) is the recognized body in New Zealand for the preparation and promulgation of national standards in all fields. Standards related to

Table 5
MEAN POWER REQUIREMENT (IN PRIMARY ENERGY TERMS) FOR ELECTRIC LIGHTING

Building type	Mean illuminance (lux)	$\bar{q}_1$ (W/m²)
Office		
5 day week	500	24.0
6 day week	500	29.0
Shop		
5 day week	500	23.0
6 day week	500	27.0
Factory (rough work)		
5 day, single shift	300	8.0
6 day, single shift	300	10.0
7 day, multiple shift	300	34.0
Factory (general)		
5 day, single shift	500	17.0
6 day, single shift	500	20.0
7 day, multiple shift	500	66.0
Factory (fine work)		
5 day, single shift	750	24.0
6 day, single shift	750	30.0
7 day, multiple shift	750	96.0
Warehouse	300	8.0
Residential	100	15.0
Hotel	100	15.0
Hospital	100	15.0
Educational	300	13.0

Note: $\bar{q}_1$ = mean power requirement (W/m²);
1 Btu ft^{-2} hr^{-1} = 3·155 W m^{-2}.

From CIBS Building Energy Code, Part 2 (Section a), Chartered Institution of Building Services, London, 1981.

Table 6
PERFORMANCE INDICATORS FOR SOME COMMON BUILDING TYPES

Building	Occupancy	Performance indicator for stated classification (GJ/m^{-2})				
		Good	Satisfactory	Fair	Poor	Very poor
Office	Single shift, 5 day week	<0.7	0.7—0.8	0.8—1.0	1.0—1.2	>1.2
Factories	Single shift, 5/6 day week	<0.8	0.8—1.0	1.0—1.2	1.2—1.5	>1.5
Warehouses	Single shift, 5/6 day week	<0.7	0.7—0.8	0.8—0.9	0.9—1.2	>1.2
Schools	Single shift, 5 day week	<0.7	0.7—0.8	0.8—1.0	1.0—1.2	>1.2
Shops	Single shift, 6 day week	<0.7	0.7—0.8	0.8—1.0	1.0—1.2	>1.2
Hotels	Continuous, 7 day week	<1.3	1.3—1.5	1.5—1.8	1.8—2.2	>2.2

Note: Single shift occupancy implies normal daily use of about 10 hr including allowances for after hours cleaning and traditional holiday periods. 1 GJ m^{-2} year^{-1} = 88,000 Btu ft^{-2}.

From CIBS Building Energy Code, Part 4, Chartered Institution of Building Services, London, 1982.

Table 7
CORRECTION FACTORS FOR THE PERFORMANCE
INDICATORS LISTED IN TABLE 6

Item	Specification	Correction factor
Occupancy[a] period	Single shift, 7 day week	1.2
	Double shift, 5 day week	1.3
	Double shift, 7 day week	1.4
	Continuous	1.4
Engineering	Air conditioning	1.4
services	Mechanical ventilation	1.3
	Electrical heating	0.8
	Purchased heat	0.8
Location	Scotland or very exposed sites	1.1
	South West England	0.9
	Remainder of U.K.	1.0

Excludes hotels — no correction factor should be applied to hotels. Single
shift occupancy implies about 10 hr occupancy per day. Double shift occupancy
implies at least 15 hr occupancy per day.

From CIBS Building Energy Code, Part 4, Chartered Institution of Building
Services, London, 1982.

buildings are prepared under the supervision of the Building and Civil Engineering Sectional
Committee of the Standards Council. The Sectional Committee includes representatives of
all the major facets of the building industry and is responsible for deciding the priorities for
standards within its field. The drafting of individual standards is assigned to a project
committee which operates by consensus.

SANZ published their "Code of Practice for Energy Conservation in Non-Residential
Buildings"[71] in 1982. The purpose of this standard was to put forward energy budget figures
for buildings and provide assistance to designers and operators wishing to meet those targets.
The underlying philosophy of the document was that buildings should be designed and
operated making the fullest possible use of all natural resources and phenomena, such as
natural ventilation and daylighting, and that the necessity for artificial cooling should be
minimized.

The standard is entirely performance oriented and defines both energy design targets (for
building designers) and energy consumption targets (for building managers).

Both space-heating and lighting energy design targets are considered. The space-heating
energy design target attempts to minimize the potential heating load due to the building
fabric. Simply stated, "The overall annual fabric heat loss should not exceed that of a cube
of the same volume and containing the equivalent floor area of the proposed building, the
same inter-floor height being maintained." The "cube" is specified in terms of its orientation, glazing, color, thermal resistance, inside temperature, location, and ventilation; and
it is recommended that the Annual Loss Factor (ALF) method[72] be used to perform the
required check calculations of annual heat losses. As far as lighting energy design targets
are concerned, designers are directed to the methods documented in Part 1 of the CIBS
Building Energy Code. The drafters of the New Zealand standard have deliberately avoided
the use of fuel factors which they see as outside its scope.

Energy consumption targets (see Table 8) are given both for existing and for new buildings.
As far as existing buildings are concerned, the figures for each activity category approximate
to the average consumption found during surveys[73,74] of around 2000 commercial and institutional buildings in New Zealand. They include all fuels and end uses. New buildings,

Table 8
ENERGY CONSUMPTION TARGETS
FOR NEW AND EXISTING BUILDINGS

Activity category	Energy consumption target kWh per m² gross floor area, per year	
	Existing buildings	**New buildings**
Personal services, con-sultants, technical serv-ices, banks, office buildings, retail trading	200	100
Restaurants	400	200
Wholesale trading	150	80
Class rooms	80	40
Industrial buildings		
With substantial process energy	100	0
With little or no process energy	200	100
Hospitals	500	250
Laboratories	200	100
Theaters	200	100
Museums, galleries	200	100
Sports buildings	200	100

Note: $1 \text{ kWh m}^{2-} \text{ year}^{-1} = 3.6 \text{ MJ m}^{-2} \text{ year}^{-1}$
$= 317 \text{ Btu ft}^{-2} \text{ year}^{-1}.$

From New Zealand Standard 4220, 1981.

taken in this context to mean those built to the energy design targets of the standard have been assigned consumption targets approximately half the average value for existing buildings. The remainder of the document (and by far the bulk if it) is taken up by recommendations concerned with the energy conscious design and management of buildings.

Like the CIBS, SANZ has attempted to put together a self-contained, performance-oriented standard. The prediction option relies on the use of a promising, but relatively untried, method for calculating space heating requirements. The energy consumption targets of the monitored option are realistically based on current practice, though perhaps set at rather more cautious levels than the CIBS. The document is recommendatory at this stage.

V. SUMMARY OF NATIONAL CONCERNS

Faced with the prospect of fuel supply shortages, many countries have been forced to examine their energy supply and demand situation. This has led to the realization that a large proportion of primary energy demand (approaching 40% in many cases) relates to the provision and maintenance of the buildings of a nation. Thus, it is hardly surprising that the appropriate governmental and professional agencies in many countries have perceived the formulation of energy performance standards for buildings as a prime concern. Attempts to tackle this concern have received considerable support from the activities of several international agencies and have coincided with a heightening interest in the application of the performance approach to building standards in general.

Since much of the energy used in buildings is concerned with the satisfaction of environ-mental requirements of one kind or another, part of this chapter was devoted to an examination of existing standards for the building environment. This examination revealed not only that energy considerations had been substantially ignored (which may not be particularly surprising), but also that many environmental requirements had been poorly specified from the human point of view and were now being substantially revised.

As far as the formulation of energy performance standards is concerned, both the prediction and the monitored type have been examined. In the case of prediction performance standards, the deficiencies of earlier "prediction" methods have been revealed, and considerable effort has been expended in the development of improved methods. In the case of monitored performance standards, the need to specify realistic target figures has been deeply felt, resulting in the conduct of surveys to improve the knowledge of energy use patterns in existing buildings.

We have deliberately not attempted a formalized comparison of the current published standards. In such a rapidly changing field, individual standards are being developed and revised at such a rate that any such comparison would become out of date almost before it could be written. However, many interesting similarities and differences have emerged between the various approaches examined. For example, a wide variety of bases have been employed for the predictive standards. An attempt has been made to root these in design practice in the case of the BEPS, while the Californians have drawn up model buildings. ASHRAE, CIBS, and SANZ, on the other hand, have proposed methods whereby each building in effect requires its target figure to be calculated. Unfortunately, the range of fuel factors incorporated, renders meaningless any direct comparison of the target figures. A different method of calculation is proposed in all five cases examined. These range from the various computer methods favored by the Americans, through the manual, but computer-based, ALF method, to the new manual method put forward by the CIBS. In a sense, these reflect national resources and expectations, both in computing technologies and in building practices.

For those groups wishing to put forward monitored performance standards, surveys of existing building energy use patterns have been seen as an essential prerequisite. In the main, the results of these surveys seem adequate for the purpose of proposing realistic targets, though there is considerable scope for judgment in setting the final figure. One minor irritant in all this is the variety of units in which targets are expressed. These range from the use of thousands of British thermal units (the so called MBtu) in the U.S. to the gigajoule in the U.K. and the kilowatt-hour in New Zealand, all normalized to a basis of floor area (in square feet or square meters). A multi-unit approach has been taken in this book to enable readers to readily comprehend what is being done in other parts of the world.

Most of these standards attempt to take account of such factors as climate, type of fuel, and extent of thermal environmental control factors which will be addressed in more detail (in Chapter 6) under the heading "Designer Concerns — Systems Energy Consumption". None of the standards dealt with the capital energy requirement of a building, an important designer concern that will be considered in Chapter 5. All of them take careful cognizance of the prime importance of the human activity occurring within the building, a matter which will now be examined in Chapter 4 under the next principal group of concerns on our agenda, those of the building owner.

REFERENCES

1. **Odum, H. T. and Odum, E. C.,** *Energy Basis for Man and Nature,* McGraw-Hill, New York, 1976.
2. *Resources and Man,* National Academy of Sciences Publ. 1703, W. H. Freeman, San Francisco, 1969.
3. **Odum, H. T.,** *Environment, Power and Society,* John Wiley & Sons, New York, 1971.
4. **Meadows, C. H., Meadows, D. L., Randers, J., and Behrens, W. W.,** *The Limits to Growth: A Report for the Club of Rome's Project on the Predicament of Mankind,* University Books, New York, 1972.
5. **Carroll, D.,** Energy consumption and conservation in buildings — an international comparison, in Proc. CIB-W67 3rd Int. Symp. Energy Conserv. Built Environ., Dublin, Ireland, 1982.
6. **Granum, H. and Hagen, H.,** Energy consumption and conservation in buildings in Norway — recent developments and trends, in Proc. CIB-W67 3rd Int. Symp. Energy Conserv. Built Environ., Dublin, Ireland, 1982, 1.A.86.
7. **Christophersen, E. and Jensen, O.,** Energy consumption and conservation in buildings — Denmark, in Proc. CIB-W67 3rd Int. Symp. Energy Conserv. Built Environ., Dublin, Ireland, 1982.
8. **Seppanen, O. A.,** Energy consumption and conservation in buildings in Finland, in Proc. CIB-W67 3rd Int. Symp. Energy Conserv. Built Environ., Dublin, Ireland, 1982, 1.A.28.
9. **Hammarsten, S. and Svensson, A.,** Energy conservation in Swedish buildings, in Proc. CIB-W67 3rd Int. Symp. Energy Conserv. Built Environ., Dublin, Ireland, 1982, 1.A.102.
10. **Koenigsberger, O., Mahoney, C., and Evans, M.,** Design of Low Cost Housing and Community Facilities, Vol. 1, United Nations Department of Social and Economic Affairs, New York, 1971.
11. Annual Report on Energy Research Development and Demonstration Activities of the IEA, 1979-1980, Organisation for Economic Cooperation and Development, 1981.
12. **Curtis, D.,** Presentation to IEA Working Party on Energy Conservation, International Energy Agency, Energy Conservation in Buildings and Community Systems Programme; Operating Agent, Oscar Faber Partnership, Marlborough House, St. Albans, Herts, England, September 21, 1981.
13. **Zegers, P.,** *Ed.,* The Communities Energy R & D Programme: Energy Conservation, 2nd updated ed., Report No EUR 7389 EN, Commission of the European Communities, Brussels, 1982.
14. About CIB, International Council for Building Research Studies and Documentation, CEP/PR 20209, Rotterdam, The Netherlands, 1981.
15. Geographical Breakdown of Membership, International Council for Building Research Studies and Documentation, Rotterdam, The Netherlands, January 1982.
16. Proc. CIB-W67 3rd Int. Symp. Energy Conserv. Built Environ., Dublin, Ireland, 1982.
17. REHVA 1980 Yearbook, Representatives of the European Heating and Ventilating Association, REHVA Clearing Centre, Oslo, Norway, 1980.
18. CLIMA 2000, Proc. 7th REHVA Int. Congr. Heating Air Conditioning, Budapest, September 17 to 19, 1980.
19. **Acton, J. P., Graubard, M. H., and Weinschrott, D. J.,** Electricity Conservation Measures in the Commercial Sector: The Los Angeles Experience, Project R-1592-FEA, Rand Corporation, Santa Monica, Calif., September 1974.
20. **Livesey, P. M.,** National savings, *J. CIBS,* 3, 8, 51, 1981.
21. **Livesey, P. M.,** PSA Energy Conservation Group Leader, personal communication, Property Services Agency, Croydon, November 1981.
22. Energy in Architecture, American Institute of Architects, Washington, D.C., 1981.
23. RICS Joins CIBS and RIBA Energy Scheme, *J. CIBS,* 3, 4, 5, 1981.
24. *Energ. Conserv. News,* 4, 8, 1982.
25. ASHRAE Standard 62-1981, Ventilation for Acceptable Indoor Air Quality, American Society of Heating, Refrigerating and Air Conditioning Engineers, Atlanta, 1981.
26. **Levy, A. W.,** The CIE visual performance system, *Lighting Res. Technol. (London),* 10, 1, 19, 1978.
27. NZS 4218P: 1977-Provisional New Zealand Standard: Minimum Thermal Insulation Requirements for Residential Buildings, Standards Association of New Zealand, Wellington, 1977.
28. ASHRAE Standard 90-75, Energy Conservation in New Building Design, The American Society of Heating Refrigeration and Air Conditioning Engineers, New York, 1975.
29. Code for Energy Conservation in New Building Construction, Natl. Conf. of States on Build. Codes and Stand., McLean, Va., December 1977.
30. Measures for Energy Conservation in New Buildings 1978, Associate Committee on the National Building Code, National Research Council of Canada, Ottawa, 1978.
31. **Dunn, R. H.,** Secretary to the Associate Committee on the National Building Code, personal communication, Ottawa, June 1978.
32. **Achenbach, P. R. and Heldenbrand, J. L.,** Development of performance-based energy conservation standards for buildings, in 1st Can. Build. Congr. — Energy and Build., Toronto, October 25 to 27, 1976, 213.

33. **Knight, J. C. and Hadley, L. G.,** Building services — an energy demand review, in Energy Conserv. Energy Manage. Build., joint conf. Br. IHVE, IES, IoF, and DoE, London, November 13 to 14, 1975, 229.
34. **Cornell, A. A., and Scanlon, P. W.,** Energy targets for office buildings and barrack blocks, *Build. Serv. Eng.,* 43, 118, 1975.
35. **Owens, P. G. T.,** Energy budgeting, *Build. Serv. Eng.,* 42, A20, 1975.
36. Report on the Seminar on Field Studies for Energy in Buildings, Coseners House, January 1980.
37. ASHRAE Standard 90A-1980: Energy Conservation in New Building Design, American Society of Heating, Refrigerating and Air Conditioning Engineers, New York, 1980.
38. ASHRAE Standard 100.1P: Energy Conservation in Existing Buildings — Low Rise Residentials, American Society of Heating, Refrigerating and Air-Conditioning Engineers, Atlanta, March 1981.
39. ASHRAE Standard 100.2P: Energy Conservation in Existing Buildings — High Rise Residentials, American Society of Heating, Refrigerating and Air-Conditioning Engineers, Atlanta, March 1981.
40. ASHRAE Standard 100.3P: Energy Conservation in Existing Buildings — Commercial, American Society of Heating, Refrigerating and Air-Conditioning Engineers, Atlanta, March 1981.
41. ASHRAE Standard 100.4P: Energy Conservation in Existing Buildings — Industrial, American Society of Heating, Refrigerating and Air-Conditioning Engineers, Atlanta, January 1981.
42. ASHRAE Standard 100.5P: Energy Conservation in Existing Buildings — Institutional, American Society of Heating, Refrigerating and Air-Conditioning Engineers, Atlanta, March 1981.
43. ASHRAE Standard 100.6P: Energy Conservation in Existing Buildings — Public Assembly, American Society of Heating, Refrigerating and Air-Conditioning Engineers, Atlanta, March 1981.
44. ASHRAE Standard 105P: Standard Methods of Measuring and Expressing Building Energy Performance, American Society of Heating, Refrigerating and Air-Conditioning Engineers, Atlanta, March 1981.
45. Energy Conservation Standards for New Buildings Act of 1976, Public Law 94-385, Title III, Washington, D.C., 1976.
46. Part 435 — energy performance standards for new buildings, *Fed. Regist.,* 44, 230, 68166, November 28, 1979.
47. Energy performance standards for new buildings: proposed rulemaking and public hearings, *Fed. Regist.,* 44, 230, 68120, November 28, 1979.
48. The Standard Evaluation Technique, Adm. Rec. No. 9561.00, U.S. Department of Energy, Washington, D.C., November 1979.
49. Statistical Analysis, Adm. Rec. No. 9562.00, U.S. Department of Energy, Washington, D.C., November 1979.
50. Energy Budget Levels Selection, Adm. Rec. No. 9563.00, U.S. Department of Energy, Washington, D.C., November 1979.
51. Weighting Factors, Adm. Rec. No. 9564.00, U.S. Department of Energy, Washington, D.C., November 1979.
52. Standard Building Operating Conditions, Adm. Rec. No. 9565.00, U.S. Department of Energy, Washington, D.C., November 1979.
53. Draft Regulating Analysis, Adm. Rec. No. 9566.00, U.S. Department of Energy, Washington, D.C., November 1979.
54. Draft Environmental Impact Statement, Adm. Rec. No. 9567.00, U.S. Department of Energy, Washington, D.C., November 1979.
55. Economic Analysis, Adm. Rec. No. 9568.00, U.S. Department of Energy, Washington, D.C., January 1980.
56. Passive and Active Solar Heating Analysis, Adm. Rec. No. 9569.00, U.S. Department of Energy, Washington, D.C., November 1979.
57. Climate Classification Analysis, Adm. Rec. No. 9570.00, U.S. Department of Energy, Washington, D.C., November 1979.
58. A first look at the proposed federal energy performance standards, *Archit. Rec.,* p. 133, February 1979.
59. **Fleming, W. S. and Misuriello, H.,** BEPS and ASHRAE 90: evaluation and impact, *ASHRAE J.,* 22, 6, 31, 1980.
60. Energy Conservation Standards for New Residential Buildings Except Apartment Houses with Four or More Habitable Stories and Hotels, Conservation Division, California Energy Commission, July 1982.
61. Energy Conservation Standards for New Nonresidential Buildings, Conservation Division, California Energy Commission, July 1982.
62. Final Report on Preliminary Analysis of, and Priorities and Required Funding for, Recommended Changes to the Title 24 Energy Conservation Standards for New Nonresidential Buildings, Energy Management Consultants Inc., Los Angeles, September 1976.
63. **Crow, L. W. and Holladay, W. L.,** California Climatic Thermal Zones Related to Energy Requirements for Heating Ventilating and Air Conditioning, California ERC&DC, Sacramento, November 1976.

64. Staff Report — Energy Conservation Standards for Nonresidential Buildings, California, ERC&DC, Sacramento, May 1977.

65. **Ayres, J. M. and Lau, H.,** Development of Building Models for Energy Budgets for Building Energy Performance Standards, plus Appendix I:1974 Building Model Specifications, Appendix II:Title 24 Building Model Specifications, Appendix III:Capital Cost Estimates and Appendix IV:Computer Program Survey, Ayres Associates, Los Angeles, October 1977.

66. Building Energy Analysis for Nine Typical Buildings, Consultants Computational Bureau, Oakland, December 1977.

67. Energy Conservation Design Manual for New Non-residential Buildings, State of California, ERC&DC, Conservation Division, Sacramento, October 1977.

68. California Lists Energy Analysis Computer Programs Approved for Conservation Standards, in Energy, the Newsletter of the AIA's Energy Notebook, Washington, D.C., June 1978, 29.

69. CIBS Building Energy Code, Chartered Institution of Building Services, London, Part 1, 1977; Part 2, Section (a), 1981; Part 3, 1979; Part 4, 1982.

70. **Stewart, L. J. and Colreavy, J. P.,** Summary of Report on Collation and Analysis of Data for the Assessment of Building Energy Targets, Building Services Research and Information Association, Bracknell, 1980.

71. NZS 4220: 1982 Code of Practice for Energy Conservation in Non-Residential Buildings, Standards Association of New Zealand, Wellington, 1982.

72. **Bassett, M. R. and Trethowen, H. A.,** ALF Design Manual: Aid to Thermal Design of Buildings, Tech. Paper P28, Building Research Association of New Zealand, Wellington, 1980.

73. **Baird, G., Donn, M. R., and Pool, F.,** Energy Demand in the Wellington Central Business District — Final Report, Publ. 77, New Zealand Energy Research and Development Committee, Auckland, 1982.

74. **Beca, Carter Hollings and Ferner and Shaw, R. A.,** Greater Auckland Commercial Sector Energy Analysis, Publ. 45, New Zealand Energy Research and Development Committee, Auckland, May 1979.

Chapter 4

OWNER CONCERNS

I. INTRODUCTION

Our first task is the clarification of the term "owner". Used here as a collective noun, "owner" refers to all those who have an interest in financing, developing, building, investing in, leasing, or even occupying a building or group of buildings. All are concerned to some degree with energy use in "their" building(s), but they all have very different objectives. We have addressed these different objectives in other chapters. By contrast, this chapter contains material of essential interest to all building owners. We use the term "owner" to refer to them all in this chapter because it is the term most likely to apply to the financier, the developer, the builder, the investor, the lessor or even the occupier. Anyone likely to have an interest in the location and size of, or the activity housed in a building is known by the one label, "owner".

As described in Chapter 2, the various decisions about its endogenous factors that the nation has made form the exogenous environment within which owners must operate. They make choices about the location of a building, the activity housed, and its size. At first glance, the factors dealt with in this chapter do not appear to be controlled greatly by the individual building owner. Hence, it might be argued that in the examination of all the influences that can potentially affect energy use in buildings, the exo-endo framework appears to be just another idealized classification system that is fine in theory but is of little practical use. This chapter sets out to prove the contrary, that there is much that may be controlled, in the location, size, and even activities housed in a building which affect the use of energy in that building. In the process the utility of the classification system will be highlighted.

Even if only one of these factors is exogenous for the owner, the data contained in this chapter will be useful. For example, local and central government can often exert more control over the size and location of a building than can the owner. Nevertheless, when selecting the scale of a new building project or when selecting new premises (from a range of existing buildings) it is essential that the energy consumption consequences of these decisions be well understood.

The research results will confirm many of the reader's common-sense beliefs. It is our contention, however, that these results can convert those common-sense beliefs into understanding based on systematic and quantitative evidence, facts that carry weight at committee or boardroom level. Readers should find the basis for sound evaluation of the energy use consequences of their decisions — facts, not guesswork!

This chapter concentrates on the decisions that can be made by the owner, but it will be clear from a brief perusal of the other chapters that they also contain information of which building owners ought to be aware. Concerned owners can influence, if not control, much that is dealt with in these other chapters. The architect and engineer will have to work within the owner's basic decisions about the location and size of a building but there are many other design concerns of which owners ought to be aware. In this chapter, we present material which can contribute towards sound decisions at boardroom or committee level, decisions which are vital to the building owner.

II. CONSTRAINTS AND CHOICES

The emphasis of this section will be placed on choices. It is useful, however, to place those choices more clearly in context by reviewing the exogenous factors that constrain building owners' choices. In the exo-endo framework, the nation, in the form of various statutory authorities, makes choices about issues which become exogenous factors for a building owner. Society, through national and international economics, sets the commodity prices which affect decisions on building materials for construction and fuel for operation. Society, through national politics, modifies prices and policies to embody national resource concerns and hence affect individual decisions about building location, preferred form of energy supply, level of insulation, and so on. These same political bodies, whether local or central, also make a large number of other rules, for reasons not connected with building energy use, which affect that energy use in many significant ways. These exogenous factors were discussed in Chapter 3 and they form the environment within which a building owner must operate.

Many would argue that building size and location plus the activities housed in a building are not factors about which an owner has much choice. They are exogenous not endogenous. Society, through individual organizations investment policies, predetermines the size of building needed to house the already well-known activities of each organization, at a location selected because of an economic or other link between those activities and a particular locality. The board of a large marketing company may believe that its head office has to be located near, and hence in the climate of, its target market, rather than in the "sunbelt" of their particular nation. The proponents of electronic communication would advocate that this is not necessary. Our argument for the value of this chapter, however, is based on the need, as we see it, to understand the consequences of making such a decision. A rational economic decision about locating the head office of the marketing company would try to balance the economic advantages of direct local access to the market against the disadvantages of providing, say, additional environmental control for a building in a harsh rather than a friendly climate.

We believe, that although all the factors examined in this chapter are to some extent exogenous, understanding their energy consequences is essential knowledge for building owners. For example, if the different AEUI ratings to be expected in different locations are known, then it becomes feasible and indeed very useful for an owner to compare building efficiencies in different locations.

We have treated the discussion of location, activity, and size as matters in which building owners have a range of options from which to choose. Readers for whom some choices are limited, should interpret the research results which deal with those choices in the light of their own needs.

The three major groups of factors — location, the activities housed, and size of a building — are treated throughout this chapter in a particular hierarchy. It is based on the degree of choice a building owner is likely to have over each of the three factors. The most exogenous of the three is location, primarily the influence of climatic factors on energy use. After this comes the activity housed in the building at the selected location. Finally there is that building factor which, of the three, is most likely to be under some degree of owner control, building size. All three of these factors can be examined and indeed measured in many different ways. The following sections are devoted to a definition of the three terms: location, activity, and size.

A. Location

The climate of an area is the prime characteristic of location that can confidently be expected to have a significant effect on building energy use. For building owners it is clear that the climate in their country is an exogenous constraint. Within certain, well-defined and documented bounds the climate of a given region follows a steady pattern. Only when a company reaches multinational proportions can it contemplate selection from among an international range of climates.

In dealing with climate in this chapter it is not intended to imply that building owners can achieve complete control of the locations of their buildings. A large number of less obvious exogenous factors to do with marketing strategies, administration, and communication practices will very severely constrain their choice in this matter. The intention here is to ensure that, in weighing up the pros and cons for decisions about location, the energy penalties of particular choices are made clear.

Investigations of the influence of climate on building energy use must start with examination of the climatic data from the nearest meteorological station. Usually the climate at the building site will differ even from that at the nearest station. Normally, an intelligent guess must be made in translating the meteorological data to the actual building location.

The macroclimate for the whole area in which the building is constructed is characterized by the data of the local meteorological station. The detailed microclimate of sun, shade, wind, and temperature at the site is less easily characterized.

The assessment of the impact of a microclimate of a site on a building, and indeed of that building on the microclimate, remains more of an art than a science. It is possible to make many generalizations about isolated buildings but almost impossible to reach similar conclusions about the microclimate around groups of buildings. It is even less clear, from research or theoretical evidence, what will be the likely impact on energy consumption of particular aspects of microclimate like solar shading or wind exposure.

1. Microclimate

There are many architectural and engineering texts that deal with this topic. They range from comprehensive meteorological treatises on micrometeorology[1] to chapters on site climate planning and its influence on design.[2-5] It is sufficient for our purposes to note the major influences on energy use in buildings and their likely variation from site to site.

The first and principal influence is the sun. It can be argued persuasively that, since it is the ultimate source of all natural energies on the earth, all that one needs to consider in natural energy flow studies is the sun: "solar-powered" temperature differences drive the wind, "solar-powered" evaporation is at the heart of rainfall and hence hydro-power, and photosynthesized solar power is at the heart of wood and given the odd 100,000 years, coal, gas, and oil power. However, in considering building energy use it is better to examine the energy as it crosses the building boundary rather than at its ultimate source. This is the level we all understand because we all experience it directly. It is the level at which we buy (gas, electricity) or use (sunlight, wind) our energy supply. All the building energy use study results in this book are presented in these terms. The energy crossing the building boundary is largely a characteristic of building or system design, while the energy needed to bring it to that boundary is much more a matter for political discussion.

Direct sunlight is of such importance at the "building boundary" that its absence can make the difference between a minimal and a high heating requirement in a cold climate. By contrast, in a climate where the air temperature alone would not make cooling necessary, the presence of the sun can make cooling essential. To a large extent this second situation can be controlled by design. The first, however, is of much greater importance as it is so little under the control of the building designer. No matter how well designed, a building on a site that is shaded from the warmth of the winter sun will have a higher than normal heating energy consumption.

It is as advantageous for the owner of a large building to think about the access to sunlight of the building as it is for the owner of the more obviously solar-heated low energy house. The scale and nature of energy need may be radically different, but the importance of making use of natural energy is still vital to the achievement of low energy consumption. The energy of the sun will be used in different ways, but it will be used. In houses in temperate climates it can be a pleasurable experience to sit in the sun. In an office in a large building this is not necessarily the case. In the former situation there is an element of choice; in a house, there are locations, other than the sunny one, to which people can retreat if conditions become unbearable. Such a choice is not normally available in a work situation. Any direct use of the sun in large buildings must allow the occupants some of the means of control that are available in a house; adjustable solar radiation levels, whether achieved with moveable sun shades or by other means, are essential.

There is another aspect of solar energy that is receiving much research attention. It is the use of daylight in the design of large low-energy-use buildings. In large buildings the energy (heat) losses through their outer surfaces (skin) are much less important to the overall energy use picture than the heat gains from artificial lighting and solar radiation. Cooling becomes an important energy load even in climates where the outside temperatures do not reach levels that would suggest it is needed for comfort. Using daylighting strategies for these buildings, while excluding the heat from direct solar radiation, can lower both the cooling load and the electrical load that otherwise would have been necessary for the artificial lighting.

Theoretical modeling indicates that in some circumstances the net gain in energy efficiency is small for both lighting and heating aspects of solar energy. Opening up the building fabric with windows to let more "light energy" into the building has the unfortunate consequence that it also increases the conduction of heat out of the building. The large "gains" through lowered need for energy to supply lighting and heating are negated to some extent by the increased energy losses through the glass. That extent can be controlled by the building owner in the selection of a site. For solar heat energy to be useful the direct radiation of the sun must be available between 9 a.m. and 3 p.m.;[6] for daylight to make a useful contribution the building, windows must have a largely unobstructed view of the sky;[7] for heat losses through the solar energy collectors (skylights, windows, etc.) to be minimized the building must occupy a sunny and sheltered site.

Other aspects of the microclimate vary with different types of macroclimate and influence comfort requirements. For example, in hot, humid climates, direct sun is a nuisance and daylighting is achieved by radically different means to those used in cooler climes; wind or air movement becomes essential for comfort in hot humid areas. This is in marked contrast to cooler climates where the wind is usually considered something to be minimized.

In order to take advantage of all these factors the building owner would do well to seek the advice of an energy professional in the selection of a site. No matter whether it is selection from among several existing buildings or the selection of a site for a future building, there is a strong interaction between site climate conditions and building energy use that ought to be considered. Knowledgeable advice on the options made available by each selection is useful input to that site selection. The matrix of Table 1 indicates the choices open to a building owner for each of a range of macroclimate types. The choices embody the dichotomy at the heart of all microclimate problems, the recognition that the very act of building has an effect on the microclimate at and around a building site. It is possible, for example, to ameliorate the disadvantages of a hot macroclimate by employing appropriate building techniques, techniques which, in such a situation, would include increasing the cooling effects of air movement at ground level in as simple a manner as possible.

B. Activity

The label "activity" here refers to the occupation(s) of the people housed in a building. The type of activity may result in labels such as "office building" being used, but, in the context of the exo-endo emphasis on people, such labels must apply to the actions and energy consumption behavior of people.

At the outset of a section on people's activities it is necessary to review two items of relevance to our discussion. The first is the clarification of the use of the term "owner". This term is used here to refer to a multitude of various functions performed by a person

Table 1
MICROCLIMATE OPTIONS FOR BUILDINGS IN FOUR SEPARATE
MACROCLIMATES[4,22]

Meteorological information	Type of macroclimate			
	Cool	Temperate	Hot-arid	Hot-humid
Solar radiance, direct	Work to maximize solar exposure of areas around buildings		Create shady areas around buildings	
	Maximum solar noon exposure; shades may be needed in summer	Maximize solar noon exposure during winter; shade in summer	Avoid direct solar gain but use reflected sunlight for lighting	
Illuminance	Daylight without solar glare brings associated heat loss	Maximize daylighting from sky without solar gain — without heat gain/loss	With small windows to avoid solar gain (overheating) look for daylighting from reflected light, especially sunlight	
Sunshine duration	Morning and afternoon glare may be a problem	Morning or afternoon glare and heat gain may be a problem	Morning and afternoon glare and overheating a potentially major problem	
Equivalent radiant temp of the sky	Another source of heat loss where cloud cover low	May be source of cooling in summer; heat loss in winter	Cooling source at night	Not normally usable
Temperature of air	Create warm, sheltered outdoor areas		Insulate against heat of air and sun and build of materials that minimize internal daily variations	Insulate against heat of air (and sun on roof)
	High insulation levels, compact building	Insulate against summer heat, winter cold; shape building to enhance energy transfer as desired		
			Create cool, shady outdoor areas	
Temperature in ground, humidity	May be warmer overall to build underground; condensation protection in construction and finishes	Underground building seldom has sufficient thermal advantage to justify cost	Smaller variation and lower overall temperatures give underground construction an attractive thermal environment	Lack of air movement opportunities (to lower humidity) normally precludes underground construction
Fog, low cloud pollution	Materials and finishes must be capable of withstanding the local air no matter what its condition			
Wind speed, direction	Avoid/protect people from and build building similarly	Avoid/protect people and building from wind chill in winter; cool breezes may be desirable at other times	Enhance air movement past water for cooling to create breezes both inside and out	Enhance any air movement; create breezes through and around buildings to lower effects of humidity
	Avoid creating dangerous external wind environment			
Rain amount, intensity	Gutter, drain size; leak protection;	Gutter, drain size; leak protection;	—	Skirt drain around building to take water away from it
Snow amount	Structural capacity; thermal resistance	Structural capacity; thermal resistance	—	—

or company at the owner level of the exo-endo classification. The second is the establishment of the bases for differentiation between activities. Even the term "building energy use" becomes misleading at an activity level of energy analysis. The building design or location may influence how much is used, but it is the people it houses and the activities they perform that actually use the energy delivered to the building. The nature of the relationship between these people and the "owner" as defined for this chapter is crucial.

If the people in the building receive the bills for, or pay the cost of the energy they use, they are obtaining feedback on their energy use decisions. They will have the opportunity to modify their energy use as a result of that feedback. The quality of feedback is dependent on many owner-selected characteristics of the building. If each activity is separately metered then at least the activity-specific energy use is being recorded separately for the people involved in that activity. Centrally supplied energy-consuming services are much less likely to be individually metered. Thus it is unlikely that individual occupants could be supplied with feedback on the energy use consequences of their actions; feedback is, at best, haphazardly applied, and is often incomplete. The following three sections consider, in turn, the owner as occupier, as investor, and as developer, and examine the role of feedback on energy use for each category.

1. The Owner-Occupier

The owner-occupier with control over all of the facilities of a building is in a unique position to influence its energy performance. If the whole building is occupied by the owner then the feedback process is simplified. The only difficulty is in establishing the degree of detail needed in that feedback.

To be useful to the owner-occupier, feedback on energy use must relate directly to the process or activity using the energy. It is not enough to know what the total energy bill for last month was; the bill must be broken down for lighting, heating, and cooling. Only such a breakdown can identify the large consuming items for which small operational changes might effect large reductions in energy use if a conservation program is planned.

Whether builder or buyer, the owner-occupier is in a unique position to observe the consequences of decisions made at an early stage of construction or purchase. Decisions which:

- Allow people to be comfortable in a commercial building after hours, without operating most of the energy-consuming services of that building
- Permit the recovery of surplus heat from one activity for the uses of others
- Maximize the beneficial climatic features of the site, reducing the requirement for heating and cooling

will all minimize energy use without reducing comfort or performance.

The owner-occupier exercises greatest control over endogenous factors when activities are known before design commences. Then, activities can be provided with separately serviced zones that operate only as required. The new building can effectively be custom built to suit the activities. Existing buildings can often be substantially modified in this way.

When the owner does not occupy the whole building, such opportunities for control of the energy interactions of activities are reduced. The remaining space in the building is typically leased to whatever activity is willing to pay the rent. Little, if any, planning time can normally be spent on special zoning or other provisions for these spaces.

2. The Owner-Investor

The approach for the owner-investor is even more dependent on the available energy use feedback. While the owner-occupier would find energy use feedback of assistance in maintaining low running costs, the owner-investor is primarily concerned with a return on investment.

It is clear that the investor with no feedback about energy costs is severely disadvantaged. To ignore the impact on energy use of activity-related decisions like metering, building zoning, appropriate and inappropriate processes, provision for occupant feedback, and so on is to invite a low return on investment.

Because endogenous activity factors for investors relate to the anticipated return on their investment, they must be aware of the energy requirements of different activities. An activity with nonstandard operating hours is going to radically alter any central energy-consuming operating costs of a service as will an activity with an abnormal process energy requirement. The first step in an investor's examination of the energy use of a building must be to obtain adequate feedback on consumption by all separate energy consumers in the building.

3. The Owner-Developer

Implicit in the label ''developer'' is the idea of temporary ownership, during the period of development. For the developer, the whole of this chapter must be seen as a plea for responsible behavior. Until building purchasers become sufficiently motivated to include energy conservation potential in their selection criteria and thus demand an energy efficient building, there is little incentive for the developer to be energy conscious in the design, construction, or redevelopment of a building. The builder or developer can make it possible for the eventual owner or occupier to conserve energy, to obtain feedback on consumption in various parts of the building, and to accommodate a varied range of activities with differing energy supply requirements. They can also make it impossible to do these things.

All the aspects of building design mentioned earlier, as owner-occupier and owner-investor concerns, are relevant to the developer. For a building to be successful as a low energy design, all these aspects must be considered.

C. Size

This third area of owner concern is one where, yet again, there is often only indirect control. The owner-occupier has a certain size of organization that must be housed; the owner-investor has a specific capital sum to put to work; the owner-developer has a particular resale expectation. These factors, more than any others, decide the physical size of the building to be bought or built. The section of this chapter, on Quantifying and Modeling Owner Options will discuss the many different ways in which size can be measured.

There are building size constraints, on building owners and designers, which are generated by external authorities, rather than the internal needs of the owners' organizations. It is extremely difficult to follow specified energy-based size recommendations when maximum size or envelope shape restrictions are imposed by city authorities. Often, additional complications are imposed in the centers of cities where economic factors usually dictate that buildings are of the maximum size allowed by the governing authority. The solution becomes one of making the maximal building work as well as possible, rather than seeking a building with optimum energy performance. Until city authorities have the time and inclination to take a more holistic view of their regulations, it appears inevitable that the size and form of city buildings will be mostly determined independently of considerations of optimum building performance.

The principal point to remember about size is its pervading influence over all other aspects of building design. It is often not made clear that to select a particular size of building almost inevitably leads to a particular level of complexity in its servicing. The small consultant's office in the suburbs may well house the same number of employees as the small office on the 20th floor of a downtown office building. However, the suburban office does not expect its building to provide a phalanx of high-speed lifts, for out-of-hours as well as normal use; nor does it expect a complex, centralized heating and ventilating system which requires all windows to be sealed against the outside environment; nor even a single, constant, high level of artificial illumination throughout the whole office which enables anyone to work anywhere in the office no matter what the requirement of their task for light. The suburban office is unlikely to expect any internal transport system whatsoever and is more likely to employ individual lights and their associated switches for each task, plus smaller scale, individually controlled heating and ventilating units (small heaters and opening windows?).

The scale of all these services depends crucially on constraints imposed on the building designer by the owner's choice of building size. The selection of the appropriate size to build or buy must be predicated on a consideration of the servicing consequences, if that selection is to include low energy use as one of its criteria.

III. QUANTIFYING AND MODELING OWNER OPTIONS

Chapter 6 contains a full discussion of the prediction models which can be used to provide a building owner or designer with estimates of building energy consumption. In this chapter we shall confine ourselves to the uses to which the results of such modeling exercises can be put. In addition we look at the ways in which the factors of location, activity, and size can be quantified. This section is introduced by noting some of the modeling problems, in particular those concerned with the activity housed by the building.

In the examination of the variation in building energy use with changes of a particular factor, the normal practice is to talk in terms of a standard building, that is, a "prototypical"[7] building. The major difficulty with such an approach is the definition of "typical". Leaving aside the questioning of the energy modeling assumptions until Chapter 6, many questions remain concerning the description of the building to be entered into the prediction model. How, for example, can a "typical" activity or size be defined for a building? What information is available on the "typical" energy requirements of processes peculiar to a particular activity? What interactions will occur between these processes and space conditioning energy needs in a building housing that activity? And again, what information is available for input to the model, about the "typical" daily and seasonal variations in the energy requirements of the activity or, more precisely, about the energy-related behavior of the people engaged in that activity.

Such considerations imply that accurate prediction of actual energy use in a building is a difficult art. The only accurate "prediction" that can be achieved frequently relies on foreknowledge of the answer sought.[8] The usefulness of modeling is seen in the insights to be gained from the relative sizes of the various inputs to the model that produce the "right" answer. Major and minor influences on energy use can be readily identified for the particular building under consideration. Where actual energy use data is not available to provide this "right" answer, and especially where the building has not been constructed (so that as-built construction data is unavailable), modeling is fraught with many more difficulties.

The conventional fall-back position is to compare building designs using the same basic input data, while varying each input parameter in turn — parametric studies. It is clearly useful for the building designer to compare, for example, window size and placement strategies in buildings of a given size. However, such one-parameter studies are suspect even when that parameter is not just at the level of window placement, but more comprehensively, considers the whole of the building fabric as a single design option. This is because the design of the thermal environment of a building cannot, indeed must not, ignore the activity to be housed in that building. There are too many dangers inherent in attempting to optimize energy use features without conducting a concurrent study of the behavioral or activity-related factors.

Data resulting from energy modeling cannot provide useful information for building owners as long as it puts aside consideration of the activity-energy use interaction. Without careful consideration of this often crucial interaction, life-cycle cost calculations have little relevance. For example, attempting to model the life-cycle cost consequences of selecting a large or a small building is of little value if the energy use prediction model does not allow for human behavior. The quantities output by the model will have little relation to the actual quantities experienced on occupation of the building.

The question of quantities also comes up in connection with the input to thermal models. In particular, in what way can location, activity, or size be characterized or quantified for a building? What are the advantages and disadvantages of the range of options available for quantifying each of these factors? The next three sections examine these questions in detail for building location (climate), activity housed by the building, and building size.

To summarize: the variation observed in building activity and size is enormous; a number of studies[9-11] indicate that building area variations for particular activities are so numerous that conventional mean values for the floor area are inappropriate (the standard deviation is often larger than the mean). As a consequence, little credence can be given to rules based on the variations predicted by parametric studies.

By contrast, in studies of actual energy use in buildings the problem lies not in obtaining accurate energy use figures but in quantifying the parameters influencing those figures. To compare the energy use in buildings of varying type, site, and size, it is necessary to be able to reduce the energy use figures to a common base — in other words to normalize them. As its simplest, normalizing is the calculation of AEUI figures for different sized buildings. This energy use per unit area figure represents energy use, size-normalized (standardized) by area. The energy use rates for buildings of different sizes can be compared using their respective size-normalized energy indexes. Chapter 2 discusses this topic in depth.

A. Climate

That climate can be measured in many different ways is obvious just from the large number of widely varying parameters by which it is measured. Many different records and statistics are used to characterize the climate of a particular area. The number of sunshine hours, the direction of the strong winter winds, the deviation of the relative humidity from set limits, or the likely occurrence of particular events (hail, lightning, etc.) are all parameters that might be wanted from a meteorological measuring station. For the study of building energy performance, interest is concentrated on the statistics required for numerical studies of building-climate interactions.

It is extremely desirable, for example, to be able to compare the performance of two buildings in different climates. To be able to normalize for the influence of climate, it is necessary to place strict requirements on the type of climate statistic used and on the way it is utilized in calculations.

Conventionally, the major requirement placed on climate statistics for use in building energy performance studies is that they be calculated relative to a human, rather than an absolute standard. For example, temperature is measured relative to the comfort of people engaged in particular activities (the more active they are the lower this temperature). Similarly, solar radiation is of interest mostly when temperatures are not within a predetermined comfort range, when its heating effect might be seen as potentially useful or potentially troublesome.

Normalizing is the most common way of allowing for the external air temperature variations of climate. These are most easily quantified in terms of a single index of the external temperature which is set in relation to the indoor comfort temperature.

Normalizing for other climate parameters, such as solar radiation and wind speed is more difficult as they have less obvious comfort levels and relationships to building energy use. In addition, a large measure of the dependence of energy use on these parameters is removed when energy use is normalized by a measure of the air temperature variation, because most of the other parameters are themselves closely related to air temperature. In most climates, for example, higher than average temperatures are associated with higher than average levels of solar radiation.

Degree-day totals are the sum, over a period, of the differences between average daytime temperatures and a set-point closely related to the comfort temperature. The set-point is normally adjusted to allow for the casual heat gains resulting from the activities housed in the building. The periods for which degree days are calculated vary; when monitoring building operation, a weekly or monthly period is used; when comparing buildings, annual totals are frequently used.

Comparison of annual energy consumption data is the principal purpose for which temperature normalization has been used in building energy management programs. Performance lines[12] and other methods for comparing energy use over a particular period with the temperature during that period are all basically methods of temperature normalization. Chapter 7 contains a detailed discussion of these methods. They provide a building operator with standardization techniques for comparing energy use during different time periods in a single building, in contrast to the building owner's interest in standardization techniques for comparing energy use in different buildings.

For the researcher into building energy performance, recorded data on energy consumption is often difficult to normalize. These difficulties are not unique to, but they are at their most complex in, climate normalization. We have already mentioned the climatic features that may influence consumption and the complexities involved in normalizing for them. For research purposes a further problem arises as a result of inadequate metering of energy supplies within buildings. If one electricity meter in a building is recording both heating and lighting, then it is very difficult to separate their energy use and hence adequately normalize the electricity consumption records for the effects of climate. Although daylight levels are partly correlated with temperature (lower daytime temperatures being associated with shorter, more cloudy days, and so on) it is not a suitable normalizing variable for the lighting energy portion of the total consumption. Hence, in this and in many similar instances, the normalization process is difficult to apply with accuracy.

It is even more difficult to see a rational quantification of the site microclimate which could usefully contribute to the normalization of climatic effects on building energy performance. Since, in the main, microclimatic effects are qualitatively rather than quantitatively known, there are no numbers readily available with which to perform the normalization. Short of putting meteorological observation stations on each building site it is not possible to obtain adequate records of microclimate data with which to do quantitative analyses. Fortunately, from the results of energy management studies, it appears that meteorological station temperature data is adequate for many purposes.

The reasons for this observed correlation between energy consumption and meteorological station temperatures can be surmised rather than proven. The temperatures measured at a building site and its nearest meteorological station are clearly subsets of the same macro-climate because they are "driven" by the same meteorological phenomena. It is unlikely that the local microclimate influences at each location would be so random that they destroyed the relationship between the driving temperature and the measured temperatures. Given these links, the site and meteorological station temperatures themselves are strongly related. This relationship is demonstrated by the observed correlation between energy consumption and meteorological station temperatures.[13]

Other parameters of climate, such as humidity, wind speed, and solar radiation are not so easily correlated with building energy consumption as is temperature. They have not, as yet, been used as normalizing variables with which to standardize energy use for variations in their theoretically not inconsiderable effects on building energy consumption.

Undoubtedly, these climatic parameters do have an effect on energy consumption. The difficulty is not in measuring that effect (by, say comparing cloudy and sunny day energy use) but in extracting, from among the many different building-climate interactions, the significance and size of their individual contributions to the observed variations in energy consumption.

B. Activity

It is in connection with "activity" that the greatest difficulty in providing suitable normalizing factors for building energy consumption is found. How can an activity be quantified in order that its energy consumption might be normalized? What values, for example, should be put on retailing and administration to normalize their respective energy uses so that they can be directly compared?

There are two basic approaches by which researchers have sidestepped, rather than tackled, this problem. One approach, which is mainly used for national energy standards and in the study of actual energy in buildings, is to split activities into a range of different categories. The American Building Energy Performance Standards (BEPS),[14] the British CIBS Building Energy Code,[15] the New Zealand Code of Practice for Energy Conservation in Nonresidential Buildings,[16] and, indeed, most standards that attempt to set guidelines or budgets for energy consumption in buildings, specify values for a range of different activities.

The other approach, often used in theoretical computer-based studies, is to limit the study to a single building type and assume that the label "office", for example, is sufficient to characterize the building housing these activities. The available data[9,10] supports the view that the definition of the "typical" office building is fraught with too many problems for it to be of much use. Given the wide variations observed in both building form and size, for a particular activity, the definition of a mean, median, or even "prototypical"[7] building for that activity is a generalization that conceals rather than reveals information.

While there is a prima facie case for considering labels like "office", "shop", and "warehouse" as sufficient consideration of activity, we believe that these labels too readily allow the concealment of the aspects of a particular activity that have major influence on energy consumption. The "typical" ("prototypical"?) office or shop is not sufficiently well defined for it to be useful in building energy analysis. It is more than mere semantics to use activity labels such as "administration" or "retailing". These labels make it clear that the activities of the people inside buildings are being classified, not some particular type of building. The basis of the concept is in the classifying of activities that occur in a building as being of a particular type ("administration", "warehouse" . . .) and then deriving the name for the building type ("shop", "office", "warehouse" . . .) from the predominant activity in the building.

The strength of the concept is also its greatest weakness. Intellectually the division is simple; practically it is often difficult to apply to real buildings, especially where more than one activity is housed. Without extensive and complex metering, the division of energy delivered to a building among the activities it houses is difficult. There is still, however, much more to be gained by facing up to this problem directly rather than avoiding it.

The advantages center principally on the emphasis placed on the importance of the people whose needs and likely behavior must be a paramount design consideration and whose demands and attitudes are essential considerations in the generation of an energy management and conservation plan.

C. Size

In direct contrast to the situations arising with activity labels, there are many possible ways of normalizing the energy consumption of a building using measures of its size.

1. Volume and Floor Area

Volume and floor area are the most obvious measures of building size. They have attracted research interest and provided more practical applications than any other measures.

An early empirical survey of energy use for "office" buildings in Philadelphia established that " . . . an Energy Use Index expressed as Btu's per gross conditioned *cubic* foot per year appears to be a more effective measure of energy consumption than Btu's per gross conditioned *square* foot per year."[17]

Our own researches indicate that there is a high degree of association between building volume and area; hence neither one can be better than the other as a normalizing variable for energy use. No significant relationship between building volume and Energy Use Index standardized by building area was found for a sample of 96 commercial buildings in the Wellington Central Business District (CBD).[9]

Little other comparative information is available on the use of volume and area as standardizing variables. In studies of actual energy use in buildings, it is apparent that use is made of whatever information on size is readily available. Because building area information is more accessible, building volume is largely ignored.

Which of the four measures of building area to use (gross, gross-conditioned, net, or net-conditioned floor area) becomes a question of data sources. It is reasonable to assume that for most buildings these four are in some constant proportion to each other. The proportions may vary with building age, but not enough to introduce systematic errors.

2. Number of Occupants

The number of people occupying a building is another simple measure of size. In our studies of energy use in commercial buildings the annual energy use per person has been a more useful parameter than annual energy use per unit floor area (AEUI). The latter is useful for comparison purposes; it is the de facto standard method of data presentation in both theoretical and empirical studies of energy use in buildings. The former measure, styled the Person Energy Use Index (PEUI), was found to be more highly associated with energy use than AEUI[9] for the commercial buildings of the Wellington CBD.

The regression analyses conducted during the Wellington CBD study were designed to search for linear relationships between energy use and each one of a set of building physical description parameters. These analyses take each variable in the set and examine it for its correlation with the dependent variable, either AEUI or PEUI. The parameter exhibiting the highest correlation with either of these is selected as parameter one. In step two the computer repeats the process, looking for correlations between the remaining parameters and any remaining variation in AEUI or PEUI not associated with parameter one. This process finishes either when all the parameters have been considered or when a particular minimum level of significance in the correlations is reached.

When nonlinear factors such as powers and inverse powers of numbers of people and area were included in two such regression processes examining first energy use per unit area and then energy use per person, a distinct symmetry in results was noted. With annual energy use per unit area as the dependent variable, the first variable included in the stepwise regression analysis was the inverse of building population; with annual energy use per person, the first variable was the inverse of building area. There is obviously some nonlinearity in the use of either building population or area as a measure of size that is best taken into account by its complementary building size measure, floor area, or building population. This is good reason to believe that the density of occupation of a building, whether measured as area per person or as a number of people per unit area, will be a useful measure of, and hence normalizing factor for, energy use in the building.

Neither of these two density factors (one is the inverse of the other) is better, per se, than building area or population as a size-normalizing factor. Although they do not figure prominently in the Wellington CBD regression studies,[9] some factor made up of a combination of building area and the inverse of building population, or of building population and the inverse of building area, does. It seems reasonable to assume that this factor, once identified clearly, will be a measure of building population density, similar to, but not the same as, number of people per unit area.

3. Hours of Occupancy

One further aspect of building population that provides a measure of size is the length of time that people occupy the building. The lengths of time premises are open or people are working are not necessarily the same as the lengths of time energy-consuming services are operated. The relationship is indirect, especially with large-scale buildings.

The heating services in a building, for example, operate over a period of time at least long enough to produce satisfactory indoor temperatures during the time the building is occupied. This can mean, in a smaller lighter mass building, that the system operates as long as people are in the building, plus approximately half an hour at the beginning of the day for warm-up; but, for the building with a high mass requiring several hours warm-up prior to occupation, it could mean that the heating system may be turned off some hours before people leave because the building cools slowly. The equation that determines the heating operating hours of a building is only partly dependent on occupation hours for the building. Other factors specific to the building design or construction are just as important.

To compare two buildings with different operating hours is like comparing apples and oranges. That building A has twice the energy use per square meter (square foot) of building B is meaningless if the two buildings are used for different lengths of time.

An activity split of the type outlined in the previous section does make allowance for some of the gross differences in operation and therefore also for the actual operating hours found for different activities. It does not allow for the differences between prosperous and stagnating businesses within an activity type; the former perhaps working overtime and the latter on short-time. Even 1 hr a day can be a quite significant proportion of the total hours of operation. The Tishman Research Corporation, in association with the firm of Syska and Hennessey,[11] is the only research team to have tackled the problem of standardizing building energy use records by a measure of the number of building operating hours. In their study of 436 New York buildings they used a term based on percentage occupation with which to normalize energy use. Built into their procedure, however, was a further size measure somewhat indirectly related to building population — the degree to which the building occupants "fill" the building.

The normalizing procedure developed for this New York study reduced the energy use for each building to an equivalent consumption based on 100% occupancy and 50 hr/week operation. The resulting index, in combination with a climate-normalizing index, proved to be very effective in sorting out the major energy use variations observed in New York buildings. The correlation studies of actual energy use were less useful than those involving normalized energy use.

4. External Surface Area

The final variable worthy of mention here is that of external surface area. Whether measured as the ratio of perimeter to floor area or as surface to volume, the external surface area variable has a definite effect on actual building energy use, though this has yet to be quantified.[9]

The energy loads imposed on buildings by the vagaries of climate are dependent on the extent to which their external surfaces are in contact with the elements. The amount of exposed glass or wall, expressed as a proportion of the occupied floor area, serves as a useful index of this exposure to the elements. The higher the glass-to-floor area ratio, for example, the more likely is it that solar heating and natural daylighting strategies will be possible, but conduction heat loss could become a problem too.

Whether a surface-to-volume, surface-to-floor area, or other standardized index of energy-related building exposure is used, no strong relationship between it and energy consumption has been reported.[9,10] As a result, no normalizing factor for exposure has been established. With the current vogue for increased exposure to better utilize the natural energies of the environment, a measure of that exposure would be useful. A first step in this direction is the Solar Load Ratio,[6] comparing the energy load to the south (north in Southern Hemisphere) window area. Current investigations into energy losses and gains as functions of the solar energy collector (window) area may well produce a useful index for the comparison of buildings with differing exposures.

For the moment, the only major research result incorporating any measure of exposure to climate is the "Skin Flux Factor" used in the Hittman study[10] of commercial buildings in Baltimore. When they found no correlation between the average (weighted) conductance of the external surfaces of these buildings and AEUI, they normalized the conductance by the gross floor area. Only for the general offices with steam heat and electric cooling was any correlation observed between EUI and the resulting skin flux factor. For these buildings, depending on the type of cooling system installed in them, between 64 and 90% of the EUI variation could be associated with (explained by?) variation in skin flux factor (see Figure 7, Chapter 6).

5. Summary

To re-cap, it is impractical to compare energy use in buildings of differing size. Standardizing for size can be, and has been, performed at several levels. The most common level is that of pure physical size, namely, the volume, floor area, or number of people housed in the building. Other levels include the hours of occupancy, the fraction of the full occupancy that is actually using the building, and the amount of external surface exposed to the environment. All these approaches produce a factor with which to alter the building energy consumption figure to produce a normalized index of performance.

When establishing size-normalized indexes of energy performance it is common to adopt the procedure used by the researchers for BEPS[14] in their computer-based studies. They established "Standard Operating Conditions"[18] for each building type. Analogous to Syska and Hennessey's standard of 50 hr/week and 100% occupancy,[11] these operating conditions specified temperatures and lighting usage profiles, as well as human occupancy densities and profiles.

Chapter 6 contains a detailed examination of the whole question of estimating (or predicting) energy use. In it, the influence, not only of location, activity, and size, but also of all the other aspects of building design that may have an influence on building energy use, is discussed.

IV. THE RESEARCH BASE

It will be obvious, from a perusal of the references cited thus far, that the research base for this section is empirical. The arguments in favor of utilizing information based on empirical data have already been propounded at the start of the previous section. It is the authors' belief that this data is the most reliable indicator of the principal determinants of energy consumption in buildings.

A proviso must be added to this introduction however. We have restricted our attention to those *published* reports based on systematic quantitative studies of energy use in a number of buildings. We can see little reason for relying on the results of haphazardly gathered case studies. While we believe the experience of the energy design and retrofit professions provides worthwhile background information, it is on the results of carefully conducted systematic surveys that we have based our conclusions.

Some of the information resulting from studies on the influence of climate, activity, and size on building energy performance has been presented in the previous section of this chapter. It is not possible to examine the potential normalizing factors for these three influences without also mentioning the researchers or others who have used them. At the very least their successes can provide us with guidance to our own use of such factors. At best, as in the case of the occupant numbers measure of building size, we acquire a useful parameter against which to measure consumption in our own building(s).

The structure of this section follows the exo-endo hierarchy. As before, the basis for this order of presentation is that it expresses a very real hierarchy of potential for action on the part of the building owner.

The reader has already been referred to Chapter 6 for information on methods for the prediction of building energy use. We recommend that conscientious owners read all of that chapter to gain a better understanding of the influence of building design on energy use. It is essential that the confusion that can occur between the effects of size and of building characteristics (e.g., central heating system) be clarified as much as possible by reference to research results.

While referring to research results it is opportune to note an important aspect of interpretation which we believe is woefully ignored in some empirical research presentations. The discernment of patterns among average energy use values is one of the important tasks of the researcher. Trends and relationships are, if anything, more important than the actual values themselves. It is these patterns which confirm or disprove the existence of an association between a building description parameter and building energy use. The interpretation of variations in averages, between groups of buildings, is not complete without consideration of the variations within each group. Often, these variations are at least as large as the average for the group. Statistical testing is needed to establish the significance of the differences in group averages relative to the large variations within each group.

If we make assumptions about the population from which our sample of buildings has been selected, then it is possible to test for significance. We frequently assume, for example, that the actual measured values of, say, AEUI, are normally distributed about the average for each group, and that each group will have the same size spread about the average. Unfortunately, it is not often that such conditions are true. As in all real world situations, however, it is possible to compromise; statistics will provide a guide even if they do not give absolutely accurate data on the significance of any observed differences.

In reality, none of the groups of buildings that are examined have normal distributions about the average value for the group. Normally distributed data gives a bell-shaped frequency distribution curve centered on the mean (see Figure 1a). Actual distributions of energy use data tend to be very skewed, as indicated in Figure 1b. They are distorted heavily towards the lower values of the parameter. In appearance, such distributions resemble bell curves with long, high parameter value "tails". The essential point to remember is that general trends in average figures, no matter how "pretty" or "logical" a picture they appear to present, must be tested to ascertain the true significance of any observed pattern.

Difficulties with interpretation can mislead the unwary, but despite this, it is our belief that the results of empirical energy performance studies can provide more insight into the truly important factors influencing energy consumption than can the simpler to execute, and more simplistic, computer-based parametric studies. The principal reason for this is that the empirical studies allow for the effects of people. Both in the operation and in the use of buildings, people and their activities are of paramount importance.

A. The Climate Dependence of Energy Use

The classical illustrations at the commencement of any dissertation on a topic such as this are of Pueblo Indian settlements in the southwestern states of the U.S. and of igloos in the northern-most areas of North America. Adaptation to both materials availability and climate reaches a certain "high" in such structures.

1. Experience in Temperate Climates

In more temperate climates the extremes are absent and the reasons for particular solutions are less obvious. Interpretation of the climate interactions of the more complex structures of today, and their associated energy systems, is even more difficult. The spur to much of

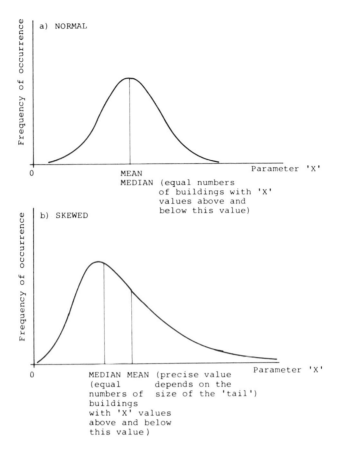

FIGURE 1.   Normal and skewed distributions of the frequency of occurrence of numbers about a mean value.

the current research into building energy performance is, in fact, the lack of an obvious or
easily understood interaction between climate and building.

The main problem is that in temperate climates (where a large proportion of the population
of the world lives) the energy costs of decisions are not urgently communicated to the
building owner/operator. The energy costs of bad building design are not great and the result
of improper application of energy design information is often just discomfort. There is no
urgent survival problem hammering home the inadequacy of the building design. The climate
of the areas where the classic igloo example originates exacts a considerably higher penalty
for bad building design.

On top of this, in our experience, people in temperate climates have a very low expectation
of the performance of their space conditioning system. They appear to have come to expect
some level of discomfort or other inadequacies in the internal environment of their buildings.
In such a context there is no pressure on the designer to improve the basic building.

The operator of the complex building is the most obvious culprit in the maloperation of
space conditioning systems. How many of you have noted, as we have, the building in a
mild climate that is operated in exactly the same manner all year round or the building,
whose heating comes on at a specific date in autumn and goes off at another date in spring
no matter what the actual outdoor temperatures are. No allowance is made for individual
differences between people or between days of the season.

As has been stated by researchers looking into passive solar energy applications in large-
scale buildings,[19] it is not possible to treat these buildings in the same way as smaller, more

domestic-scale ones. In design as well as in operation, the large building has completely different energy characteristics. Its energy use is usually dominated by the influence of internal heat gains, rather than energy transfers through its external skin.

2. The Results of Field Surveys

The results of studies in Wellington[9] and New York[11] show that in large, air-conditioned buildings, 35 to 40% of the total energy use is for heating; domestic-scale buildings devote a substantially higher proportion of their energy use to heating. The cooling load in domestic-scale buildings is purely climate dependent, while in the large office building, it is as much due to the heat gains from the lights and from the occupants.

Our own studies of the Wellington CBD now cover 4 years of energy use information. We have not yet been able to establish a link between the climate for those years and the average energy use figures. Our survey of a large number of buildings has revealed more about the variations from one utility meter reading to another than it has about variations from one year to another, as a function of climate. Comparisons from year to year have been made, but these relate chiefly to nonclimate factors such as pricing and rationing, that is, to exogenous factors for the building owner.

With empirical studies the effects of climate on energy use are not easily determined, especially if there is a large baseload of nonclimate dependent energy use. The presence of uncontrolled and unidentified influences on this baseload adds "noise" to the energy consumption data, which tends to mask any climatic effects. The choice of climatic parameters to provide the independent variables for the analysis is also important. Average dry bulb temperatures[12] and degree days[20] are commonly used but both have drawbacks. Degree days are based on an assumed cut-off temperature for heating. This cut-off temperature can easily be determined for an individual building, but for a sample of buildings no single cut-off temperature will generally exist. In addition, the success of using degree day-based predictions of energy use relies on some sweeping assumptions about the heat gains to the building. In theory, average temperature is not as directly related to heating requirements as degree days; it is not as likely as degree days to recognize the daily variations about the desired indoor temperature. In practice, this effect is small and average temperatures were chosen for our analysis of the Wellington CBD.

3. The Effect of Energy-Consuming Systems

A further difficulty was posed by the many different types of space conditioning systems installed in the buildings of the CBD. It will be appreciated that an average energy use figure covering all the system types is unlikely to bear a strong relationship to climate. The Hittman study found that the 23% difference in average cooling loads between Philadelphia and Baltimore corresponded to a 23% difference in cooling degree days (a measure of the need for cooling in a particular climate) for the two cities.[10] The need for heating (as measured by heating degree days) and the average heating fuel use figures were approximately equal in the two cities. Where consumption figures cannot be so neatly ascribed to their final uses, the derivation of a direct relationship between energy consumption and climate is difficult.

For four buildings in the Wellington CBD with large annual gas consumption but small baseloads (as indicated by their summer gas use), the monthly gas use was compared with the average, working day, outdoor, dry-bulb temperature. The energy use was shown to drop between 4.9 and 5.4% for every degree Celsius (1.8° Fahrenheit) rise in temperature between 8°C (46.4°F) and 18°C (64.4°F). The strength of the relationship indicated that any additional effects on monthly energy use, associated with other climatic parameters, were not very important.

The increased variability and reduced strength of the relationship for weekly as opposed to monthly readings in the same buildings indicated that solar radiation and wind contribute significantly to heating load variations over these shorter time periods. The effect of taking monthly readings of energy use and temperature is obviously to "average out" the effects of sun and wind on the energy requirements of the buildings.

The influence of outdoor temperature on electricity and gas consumption in the whole of the Wellington CBD was also examined. Both average values of the individual energy consumptions over all of the buildings and the total consumption for all the buildings were studied (see Figure 3, Chapter 6).

The average CBD electricity meter has an annual baseload amounting to three quarters of its total annual consumption. While its consumption does vary with temperature (from summer to winter) the variation is small, 0.5 to 1.0% of the total annual energy consumption per degree Celsius (1.8°F) decrease in temperature. The total electricity consumption for the CBD did not vary with varying temperature. It was concluded that the larger electricity consumers, whose energy use inevitably dominates the total, use energy for purposes that are independent of temperature.

For gas there is no such obvious difference between large and small consumers (as there was for electricity). The total and the average individual gas energy use figures are both dependent on outdoor temperature. The temperature dependence of individual consumption is stronger than for electricity — for each 1°C (1.8°F) drop in temperature, a corresponding 1.0 to 1.2% increase was observed in the monthly proportion of the total annual gas consumption (see Chapter 6, Figure 3).

B. The Activity Dependence of Energy Use

It is agreed[10,14,15] that the activity a building houses has a major effect on its energy use. The very fact that researchers examine offices as separate entitites from houses is a positive statement of this. There have, to our knowledge, been very few studies into the energy use values typical of different commercial activities.

As indicated in an earlier section, the approach in these studies has been to establish baseline values for each activity in the country or the city under investigation. Methods have not been established which would enable calculation of the change in energy use resulting from a change in the activity housed in a particular building.

1. The Classification of Activities

The principal difficulty facing the researcher making an empirical study of the influence of "activity" on energy use in buildings is that of access to appropriate information. At the outset, it is necessary to define a set of activity labels, which recognize the potential influence of different activities on energy consumption. "Typical office buildings", as such, do not exist. What does exist is a large variety of buildings in which people engage in "administrative" or "office" activities; the people-based behavior label is essential.

There is a second related aspect of this difficulty which must be considered in empirical studies of building energy use. Most commercial buildings house more than one type of activity, making it difficult to define a single activity label which can be used to describe a building. As a result, the building description assumes a lesser, and the activity description a greater, significance. Instead of looking for office buildings, it becomes necessary to look at administrative activities, examining the building type which they occupy. An empirical study must seek to examine particular activities, no matter where they occur, in buildings of many types. The study narrows from the whole building to a single room or floor within the building. An information problem arises because it is difficult to obtain data on the consumption of all premises within buildings; few buildings are metered in sufficient detail or in a way that separates the end uses which are to be studied.

In the commercial sector, this activity-based classification system raises yet another problem of definition: to what degree should the myriad human occupations be grouped under

Table 2
COMPARATIVE LIST OF ACTIVITY CLASSES USED IN THREE STUDIES OF COMMERCIAL SECTOR ENERGY CONSUMPTION

Auckland (N.Z.)[21]	Wellington (N.Z.)[9]	Baltimore (U.S.)[10]
Offices	Consultants, Banks, Administration (private, public), Medical services	Offices with CDS[a], Banks, General offices
Retail/services	Restaurants, Personal services, Technical services, Retail trading, Educational services	Restaurants, Personal services, Food stores, Department stores, Drug stores, Small stores
Educational	—	—
Health/Security (24-hr operation)	—	—
Recreational	Social, cultural, and recreational; Theaters, Nightclubs, and taverns	Theaters, Nightclubs
Religious/community	—	—
Commercial	Commercial	Hotels
Accommodation	Accommodation	
Miscellaneous	Wholesale trading	Stores with warehouses

a   CDS, Presence of computers, data processing equipment, and support facilities.

generic labels? It is obviously not enough to operate the de facto definition of the commercial sector: "if it's not industrial, agricultural, or domestic then it must be commercial." Everything from the general hospital to the corner store fits that definition. Conversely, to concentrate on the eating functions, the sleeping functions, the walking functions, and so on of the human person is just as obviously too fine a definition of activity.

For the studies of the Wellington CBD the activity-based divisions contained in Table 2 were made. The general philosophy governing construction of the list was a desire to isolate activity types which might have differing energy consumption patterns. For this reason traditional building-oriented labels like "shop", "office", and "warehouse" were avoided where possible.

Table 2 contains two further lists similar to that devised for the Wellington study. One was constructed for the study of Hittman Associates of Baltimore and Denver;[10] the other was constructed for a study of the commercial sector of Auckland,[21] the largest urban area in New Zealand. The similarities are highlighted by the construction and layout of the table. It is clear that simliar analyses of the commercial sector have occurred in all three studies. All have identified the importance of behavior, the occupations of the people housed in the buildings under study. For example, it was not enough for the Hittman study to identify offices; they split these offices into those where people use computers and those where they do not.

2. The Results of Empirical Studies

A summary of the results from these empirical studies is presented in Figure 2. The activities are shown in order of decreasing intensity of energy demand. As always with histograms of this type, which are based on average figures, warnings must be made about

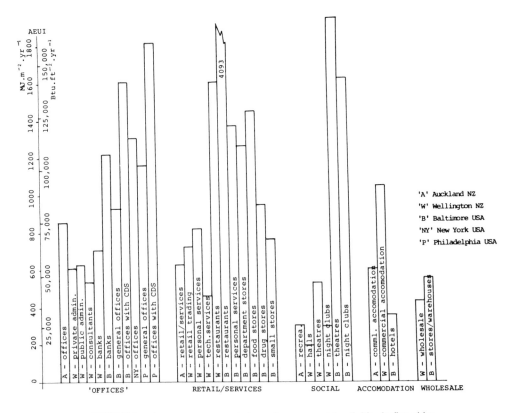

FIGURE 2.   Average energy use indexes for commercial activities in five cities.

the derivation of those averages. For example, two sets of figures can have approximately the same average value, even though one set has twice the range of the other. The conclusion that may be drawn from similarity in such a situation is limited.

In Figure 2 there is a further complication, created by the use of a different calculation method by Hittman Associates. Instead of calculating a mean value for all the individual AEUIs of all the individual premises in their study they calculated a single AEUI for all the "restaurant" or all the "general office" floor space. Thus instead of calculating the average AEUI for the 13 banks of their study they divided the total of all the annual energy use of the 13 banks by their total floor area to obtain an AEUI "typical" of the whole area. Dependent on the relationship of the larger of the energy consumers' AEUIs to the "true" mean this latter AEUI which is typical of all the bank floor space collectively, is larger or smaller than the true mean. The pattern is not consistent. For the activities with lower average AEUIs it appears that the larger consumers have lower than average AEUIs; for activities with higher average AEUIs the situation reverses; larger premises use more, not less, energy per square meter (foot) than their smaller counterparts. Fortunately it is possible to use the published[10] raw data to recalculate AEUI values for the Hittman study. These recalculated values were used in the construction of Figure 2.

It can be argued, from the wide variations in building design and construction, HVAC system type, and even AEUI found within each activity category, that activity is not the sole determinant of building energy use. This may well be the case, but the results from the Wellington CBD study have shown that activity is the prime determinant of energy use. During the analysis phase of this study a number of individual analyses were conducted. The results of these analyses, performed for six different groups, were unique to each group. They bore no resemblance to a similar analysis performed on all premises in the study.

The important features of these analyses can be summarized as follows:

1. Very different factors were identified as important determinants of energy use for each activity. Very few were important for all activities. Activity was shown by these analyses to be a far more important determinant of energy use than many building characteristics. While not of itself being the cause, activity does determine the type and size of premises, the operating hours, the type of space conditioning required, plus the requirement for other energy consuming systems; the factors that can be seen to have a direct causal connection to energy use.

2. In addition to area factors, there are other size factors related to the space allocation per person, which are of great importance in the estimation of building energy use intensity.

3. No one factor, of those used to characterize building operating hours, was found to be consistently important, although the AEUI for each activity group was related significantly to at least one of the factors.

4. Both gas- and oil-fueled buildings are associated with increased energy consumption. There are at least two likely reasons for this. The first is that buildings in the CBD using these fuels tended to be larger, more complex, and centrally serviced, an ideal recipe for increased energy use. The second likely reason is because the analysis only examined site energy use, the energy content of the electricity, gas, or oil delivered to the building. No account is taken of the efficiencies of the systems using the fuels.

5. In the case of electric lighting, buildings where the premises were metered and billed directly for their electricity use had lower AEUIs than buildings where it was centrally metered. In the case of electric power (for lift motors, chillers, fans, pumps, etc.) the situation was the reverse. Premises in buildings where the power bill was received by the building had lower AEUIs for most activities. The former result was as expected; the buildings with better consumer feedback used less energy. In the latter case, two interpretations are possible. The first is the possibility that other factors, uncontrolled and outside the analysis, were influencing these results. Principal among these is the arrangements for metering. Buildings where premises are billed for most of the electric power usually contain more meters. Such buildings could be expected to be more complex in other ways as well and as a result, the AEUI would be higher than for less complex buildings, despite better energy use feedback. The alternative interpretation is that electric power consumption is lower where a single person, the building owner, has more control over its use.

6. Increasing complexity of energy-consuming services leads to increasing (normalized) premises energy use.

These factors, while identified as more influential than all the others tested, cannot be said to determine all the variation in AEUI. There is still a high degree of unaccounted-for variability.

3. Unaccounted-For Variability

It is our belief that the remaining variability is due to those less-easy-to-quantify parameters, which identify different aspects of the management of a building and its systems. This belief has been supported by further research into individual building energy conservation programs.[13] Chapter 7 discusses these points in detail.

As mentioned in a much earlier part of this chapter, an attempt to draw conclusions about the absolute value of the impact of activity on building energy use, from the data currently available, would be difficult to justify. The information of Figure 2 can be supplemented by the data of Table 3 (in Chapter 3) which represents the output from the U.S. BEPS development program. Further than this it is hard to imagine more precision being obtained. The indications from the regression analyses in the Wellington CBD study are that, even with a good knowledge of the physical characteristics of the building housing an activity, its actual energy use can only be predicted within error limits equal to half the predicted value. Chapter 6 discusses this in detail. Here it is sufficient to recognize the importance of activity.

The first step in an energy performance analysis is, as discussed in Chapter 2, to ensure that the amenity provision of a building is similar to that in the building (or building model) providing the performance measure or yardstick. To compare a single-family house and a multistory office building is patently ridiculous; one provides domestic-scale services, the other commercial. The whole of the analysis presented here argues for a careful study of activities at an even more detailed level than the "domestic" or the "commercial". When analyzing the energy performance of a commercial building a yardstick suited to the particular class of commercial activity it houses is essential. The data presented in this section of Chapter 4 and in Chapter 3 help in the selection of, and in the determination of, suitable yardsticks of performance for the appropriate class of activity.

C. The Size Dependence of Energy Use

The first aspect we must deal with when studying the influence of the size of a building on energy use is the obvious direct deterministic connection between size and energy use. At its simplest it is the observation that if more people are doing an activity then they will require more energy. One person with one light becomes ten people with ten lights, and so on. However, no rationale can be made, as it can for production lines, that there is greater efficiency in numbers; that ten people should only require eight lights (say). Conversely, there is no case for the larger building requiring any more than the standard energy use per person.

1. Floor Area and Number of Occupants

As discussed in Chapter 2, there is some doubt as to which is the best factor to use to standardize building energy use. For example, the greater the floor area, the more energy the building requires, generally in proportion to the increase in area. The number of people in the building provides a similar measure. In the Wellington CBD study,[9] the number of people, rather than floor area, was found to be a slightly more influential determinant of energy use in buildings.

The regression analyses for Wellington CBD buildings indicate that AEUI (annual energy use divided by floor area) is most associated with the variable expressing the number of people in buildings. In the reverse situation, for PEUI (energy use divided by the number of people in the building) the regression studies indicate that the area variable is strongly associated with this PEUI. It is clear that neither measure of size, floor area, or number of building occupants is a complete, single measure. Some combination of the two is needed. It would appear from the research results that the number of people housed in a given area in a building, or conversely the area allowed per person, might be more useful standardizing variables than either area or number of occupants used individually.

The question of the direct dependence of energy use, on the area allowed per person or its inverse, is unresolved. It may be reasonable to presume that the amount of energy used in a building will increase if the number of people housed per square meter increases, but whether it increases by some direct multiple of this number, or by some amount more obscurely dependent on it, is yet to be explored.

2. Building Volume and Height

Volume is the other major measure of the size of a building which has been subjected to some investigation. In Philadelphia,[17] the gross volume proved a more satisfactory standardizing factor than gross floor area. Other studies[9,10] have found volume of less utility as a normalizing variable. It is only possible to speculate as to the reasons for this. However, it seems likely that problems of definition are at the heart of this situation. Volume is not necessarily worse than area or occupant number as a size estimation, but it is not so readily calculable at the level of accuracy needed for it to be useful in most survey work.

A relationship similar to that held by building area and number of occupants has been found between building height and energy use. The connection is directly one of size and can be removed by normalizing using other size factors. Thus, height for small stores in Baltimore[10] was found to be unconnected to AEUI. There were, it was concluded, other ''major parameters which have far greater influence on the EUI than the effect height has.''

There is no evidence that height is a better measure of size than the area of a building or number of occupants. In addition, there is no evidence of a second order effect; increasing height, while highly associated with increasing total energy use in a building, is not associated with increasing intensity of use (AEUI).

3. Operating Hours

As would be expected, the hours of operation of major pieces of equipment are very highly associated with both total and standardized energy consumption figures for a building. A building operated for 40 hr/week has a considerably smaller AEUI than one operated for 50. The New York study by Syska and Hennessy[11] attempted to take account of operating hour variations by developing a standardizing factor for it. The basis of this combined occupancy/utilization index was a simple multiplicative adjustment of the AEUI. It was assumed that, unoccupied, the building would consume one third of its normal weekly consumption and hence this index could only operate on the other two thirds. In the correlation studies of 44 buildings that are presented in the Appendices to their report, it is clear that normalization by the occupancy/utilization factor has been useful. The graphs, where the points have been normalized, have less scatter than those where they have not. Normalization has clarified those trends which exist (in relationships between type of perimeter heating system and energy use per square meter) by removing the random variations due to variations in the hours of use of these systems.

These findings have been supported by the results of the Wellington CBD study.[9] There, the most highly influential factor, after size, was the hours of use of the building. It is interesting to note that the importance of operating hours dropped considerably in the correlation studies of individual activities. For restaurants, for example, the multiple regression analysis brought out only one operating hours factor and that was ranked tenth. The factor was the hours of operation of the cleaners!

It would seem that, for many activities, definition of the activity is itself a sufficient definition of operating hours. Further attempts to account for operating hour variations become unnecessary. This argument augers well for those attempting to draft building energy performance standards; it suggests that energy use targets, based on actual consumption figures for particular activity types, are sufficient in themselves. The basic AEUI figure need not always be complicated by extra factors and allowances for nonstandard operation.

The data presented here is based on the experience of many building owners and users; generalization from others' experience is the cornerstone of the information in this chapter and is, we believe, the cornerstone of any energy conservation program for buildings.

V. IMPLICATIONS FOR OWNERS

The principal message of this chapter is that owners must make themselves aware of building energy performance whether they invest in, occupy, or intend to develop a building.

If energy prices realistically reflect the resource conservation intentions of a nation, building owners will be able to reduce their running costs significantly, quite apart from any obligations they may feel towards the conservation of national resources.

The first factor considered in this chapter was the location of a building. It is recognized that selecting the location can have a marked effect on the component materials, indeed all the resources, that are available and required for constructing and operating a building. However, we have not considered these questions directly in this chapter. The principal location parameter examined has been that of the climate. Chapter 5 addresses the whole question of how one goes about analyzing the investment of energy that is needed to make a particular component or building. These are the methods that would have to be applied for each new location one wished to examine. There is some additional information, in Chapter 6, on the operating energy consumption implications of the use of different types of construction materials. To the extent that the availability of different construction materials depends on location, this type of analysis can prove useful in assessing the energy cost in use implications of the location.

The second factor examined in this chapter was activity. The message here is very simple; there is no blanket energy use budget or target figure which is applicable to all buildings. Every building, as well as being unique in its physical features, has a different mix of activities from every other building. Factors which are important determinants of energy use for one activity may have no significance whatsoever for another. Again, building owners, no matter what their role in the day-to-day running of the building, must understand the implications for energy use of the types of activity housed in their building. The services provided for a group of consultants, who have irregular hours and an intermittent demand on office services, are very different from those provided for a group of clerical staff working regular hours. Chapter 6 contains a fuller discussion of the influence of building services on building energy performance.

The influence of building size on energy performance was the third and last question addressed in this chapter. The case was argued that there are very many ways in which one can measure the size of a building, and no particular one is of greater relevance than any other. The hours of use of a building (the size of its usage) is just as relevant to the energy performance, as the conditioned floor area within the building. Each has a complementary influence on energy use and their combined effects must be understood and accounted for in generally applicable indexes of energy performance.

The last major conclusion of this chapter was that the building user is of paramount importance. As will be confirmed by the research results discussed in Chapters 6 and 7, only some of the observed variation in recorded energy consumption can be attributed to the design of the building and its services. The rest, some 50% of the total, is attributable to the building user or operator.

Chapter 6 discusses the influence of the physical features of a building in much greater detail; Chapter 7 reports on techniques for examining and analyzing energy use in individual buildings. Before going on to consider these, however, we shall deal with the first of our designer concerns, capital energy requirements.

REFERENCES

1. **Geiger, R.,** *Climate Near the Ground,* Harvard University Press, Harvard, Mass., 1965.
2. **Oke, T. R.,** *Boundary Layer Climates,* Methuen & Co. Ltd., London, 1978.
3. **Konya, A.,** *Design Primer for Hot Climates,* Architectural Press Ltd., London, 1980.
4. **Olgyay, V.,** *Design with Climate — Bioclimate Approach to Architectural Regionalism,* Princeton University Press, Princeton, N. J., 1963.
5. **Markus, T. A. and Morris, E. N.,** *Buildings, Climate and Energy,* Pitman, London, 1980.
6. **Balcomb, J. D., et al.,** Passive Solar Design Handbook, VII, Passive Solar Design Analysis, U.S. DOE Report No. DOE-CS-0127/2, January 1980.
7. Project Independence Blueprint, Task Force Report, Residential and Commercial Energy Use Patterns 1970—1990, U.S. Government Publ. PB448-508, prepared by A. D. Little Inc., November 1974.
8. **Kerrisk, J. F., Hunn, B. D., et al.,** Custom Weighting-Factor Method for Thermal-Load Calculations in the DOE-2 Computer Program, Rep. No. LA-UR-81-1559, Los Alamos Scientific Lab., N.M., available from U.S. Department of Commerce, Washington, D.C., 1981.
9. **Baird, G., Donn, M. R., and Pool, F.,** Energy Demand in the Wellington Central Business District — Final Report, Publ. No. 77, New Zealand Energy Research and Development Committee, Auckland, New Zealand, 1982.
10. **McCarthy, P. M., Patel, R. F., and Karpay, B.,** Empirical and Simulation Analysis of Energy Use in Commercial Buildings, HIT-664-1, Hittman Associates, Columbia, Md., February 1977.
11. Syska and Hennessy and Tishman Research Corporation, Energy Conservation in Existing Office Buildings, Phase 1 Report, Department of Energy, Washington, D.C., June 1977.
12. PSA, Monitoring fuel consumption of heating installations, in Technical Instruction (M&E), Department of the Environment, Property Services Agency, Croydon, U.K., December 1974.
13. **Baird, G. and Brander, W. D. S.,** Energy Conservation in Government Buildings — Report on Stage 2, Contract Research Paper 15, School of Architecture, Victoria University of Wellington, New Zealand, February 1982.
14. Energy performance standards for new buildings, notice of proposed rulemaking, *Fed. Regist.,* 44, 230, 68120, November 28, 1979.
15. CIBS Building Energy Code, The Chartered Institute of Building Services, London, Part 1, 1977; Part 2, Section (a), 1981; Part 3, 1979, Part 4, 1982.
16. NZS 4220: 1982 Code of Practice for Energy Conservation in Non-Residential Buildings, Standards Association of New Zealand, Wellington, 1982.
17. Enviro-Management and Research, Evaluation of Building Characteristics Relative to Energy Consumption in Office Buildings, Washington, D.C., September 1975.
18. Proposed Energy-Performance Standards for New Buildings: Standard Building Operating Conditions, Rep. No. DOE/CS-0118, Department of Energy, Washington, D.C., November 1979.
19. **Ternoey, S. E.,** Don't trust your instincts: big building design defies intuition, *Solar Age,* October 1981.
20. **Mayer, L. S. and Benjamini, Y.,** Modelling residential demand for natural gas as a function of the coldness of the month, *Energ. Build.,* 1, 3, 301, 1978.
21. Beca, Carter Hollings and Ferner, and R. A. Shaw, Greater Auckland Commercial Sector Energy Analysis, Rep. no. 45, New Zealand Energy Research and Development Committee, Auckland, May 1979.
22. **Lacy, R. E.,** Survey of Meteorological Information for Architecture and Building, Building Research Station, Current Paper 5/72, Building Research Station, Garston, U.K., 1972.

Chapter 5

DESIGNER CONCERNS — CAPITAL ENERGY REQUIREMENTS

I. INTRODUCTION

Chapter 4 considered factors that are mostly within the realm of the building owner. This chapter and the one that follows will examine some of the factors that are the concern of the building designer. We have identified two main classes of concern, those related to the energy consumed by the building in use and those related to the energy required to construct the building. The former will be discussed in Chapter 6. In this chapter we shall consider the capital energy requirements of buildings.

Since the idea of capital energy requirements is relatively unfamiliar in the field of building we shall start by illustrating its relevance to energy conservation. Unless otherwise compelled to do so, designers would not normally consider capital energy requirements in their choice of building materials or components; decisions will normally be based on other considerations, most notably cost. Frequently, the costs of materials and components will reflect their energy requirements, but if this were not the case the situation could arise where the operating energy savings resulting from the use of a particular item (rather than another), turned out to be less than the (additional) capital energy requirement resulting from the use of that particular item.

For example, it has been estimated that the capital energy requirement of insulation for New Zealand houses would be balanced by the savings in operating energy over a relatively short period (2 to 5 years); the cost payback time would be of similar duration. In the case of double glazing, on the other hand, the period needed to balance the capital energy requirement could exceed the useful life of the building, even though the cost payback time is very much less. In this latter type of case, national energy planners may wish to regulate the choices of materials or components in order, for instance, to reduce the demand for a particular form of energy (which may be an essential input to the manufacturing process for a given material) or simply to reduce energy demand in general. Hence, this chapter is of relevance at the national level as well as to building designers.

The concept of embodied energy is considered first. The terminology and conventions in energy analysis and its limitations are also considered. Following that, the energy consumed by the construction industries of two countries, the U.S. and New Zealand, are examined and compared.

The processes involved and the energy consumed in the manufacture of some common building materials is then studied and the energy coefficients of these building materials are presented. The application of these energy coefficients, to the estimation of the energy requirements of building components and of a standard house, is then illustrated. The chapter ends with a brief outline of possible design applications and opportunities for the use of energy analysis.

II. THE CONCEPT OF EMBODIED ENERGY

To embody is to "cause to become part of body", or to incorporate or invest. The embodied energy of a product is the sum total of all the energy incorporated or invested, throughout the various stages of manufacture, from raw material to finished product. This section will outline the methods and conventions for calculating embodied energy. The limitations of the concept in relation to human labor and to the indirect costs associated with environmental damage are also discussed.

A. Introduction to Energy Analysis

Energy Analysis is a "formalized" method for calculating the overall amount of energy required to produce goods and services. These energy requirements include both the energy consumed directly by the particular process in question (the direct energy requirement) and the energy used indirectly to extract the raw materials associated with the process (the indirect energy).

Thus, the energy required to build a house includes both direct and indirect energy. The direct energy is the fossil fuel and electricity used at the building site and in transporting the materials to the site. The indirect energy is that used to manufacture materials, components, tools, and machinery, plus the energy required to extract the raw materials, plus the energy used to manufacture the equipment used to extract these raw materials, and so on.

There are many methods and conventions for calculating the energy requirements of a product or service, all of which give very different results. There are, however, some widely accepted terminologies, methods, and conventions. Those established by the workshops of the International Federation of Institutes for Advanced Study (IFIAS)[1,2] are the most widely adopted and will be used here. They provide a coherent framework for carrying out the calculations involved in energy analysis and for comparing the works of different authors. The methods and conventions are summarized in this section, the terminology is given in Table 1.

B. Terminology and Conventions

Many authors have found the equating of energy to enthalpy or "heat content" adequate for energy analysis. However, to analyze the production of goods and services in terms of heat can be very misleading in certain circumstances. The 1974 IFIAS Workshop[1] examined "enthalpy" and "free energy", among others, as units of energy, and recommended the following:

"Where Energy Analysis is concerned with depletion of the resource base, all figures should be expressed in terms of free energy. However, recognizing that in many cases it is not possible to compute the free energy changes of actual processes, it is sufficiently accurate in the case of intensive fuels to express figures in terms of gross enthalpy."

The first law of thermodynamics states that energy can neither be created nor destroyed. Stated another way, the total energy of a system and its surroundings remains constant. Thus, when a fuel is burnt its energy is not lost. It is merely converted to another form. However, the free energy, that is the amount actually available to do useful work, diminishes every time a fuel is burnt and is not recoverable. Enthalpy, which is the heat or energy content of a system, is not always a good indicator of the "usefulness" of energy. This is the main reason for the IFIAS Workshop recommendation that free energy, rather than enthalpy, best meets the objectives of Energy Analysis. For intensive fuels, such as oil, gas, and coal, however, the error in using enthalpy rather than free energy is only approximately 10%.

C. System Boundary

In energy analysis it is often only possible (or necessary) to calculate the process energy requirement of a product. In such cases it is necessary to draw a boundary between energy supply and demand. The actual boundary selected depends mainly on the overall aims of the particular analysis and the assumptions that are made, and partly on the availability of data. The 1974 IFIAS Workshop[1] suggested the four boundary levels described in Table 2. As indicated on the table, decreasing increments of energy are involved as the boundary increases from Level 1 to Level 4.

Table 1
THE TERMINOLOGY OF ENERGY ANALYSIS

Sequestered or embodied energy	To sequester is to store or put aside; the sequestered or embodied energy of an article is thus the energy accumulated by all the processes required to extract the raw material through to making the finished article
Energy coefficient	This is the energy required to produce a unit of material; energy requirements per unit mass or energy requirements per unit volume are two examples of energy coefficients
Direct energy requirement	This is the energy required to carry out a given process; the direct energy requirement for manufacture of cement, for example, will include only the energy used in the manufacturing process; the energy required to extract and transport the raw materials will not be included
Indirect energy requirement	This is the energy embodied in raw materials and capital, together with that required for transport and for the administration involved in a given process; the indirect energy requirements for cement manufacture, for example, includes the energy required for the extraction and transport of raw materials, the energy for making the machinery and plant (for both the manufacturing and the extraction processes), and the energy required to support administrative services
Gross energy requirement (GER)	This is the overall total energy sequestered by the process of making a product or service; it consists of the sum of all the embodied (direct and indirect) energy requirements
Process energy requirement (PER)	This is the energy involved in the manufacturing process itself; to obtain the PER, it is only necessary to know the direct energy delivered to the process
Intensive fuels	Fuels with a high free energy potential per unit mass such as oil, gas, and coal

Table 2
IFIAS SYSTEM BOUNDARIES

Level	Percentage GER[a]	Boundary
1	Less than 50%	Direct energy involved in the process only
2	Approximately 40%	Energy involved in extracting materials
3	Rarely greater than 10%	Energy needed to make capital equipment
4	Very low	Energy to make the machines that make the capital equipment

[a] Typical amount of total GER. This is only an approximate guide from the 1975 IFIAS Workshop.[2]

D. Labor Content

Due to the labor intensiveness of the building process, human labor is an important element in any analysis of energy requirements for construction. An energy-intensive process can often be substituted in part by human labor and vice versa. Considerable controversy, however, still exists as to what part of the energy support of humans as consumers should be included in the calculation of energy requirements. Odum[3] suggests that "because of its high quality and thus high energy cost, human service is the major part of any energy analysis and cannot be omitted.'' The 1974 IFIAS Workshop,[1] however, recommended that ''where the analysis refers to developed or industrialized economies it is not necessary to consider the energy for life-support or manpower.'' The method for assessing the contribution of human labor and what to include has not yet been satisfactorily resolved. Therefore, most energy analysts choose to exclude it from their calculations.

Stein et. al.[4] have calculated the labor embodied in building materials separately from the embodied energy. The embodied labor or labor requirements of different building materials, components, and assemblies can in this way be compared, as it can for the embodied energy (and correlated to it).

E. Environmental Effects

Mining and quarrying disrupt underground water circulation, damage landscape, and induce soil run-off, thereby reducing the productive agricultural land area. Hydroelectric power dams can reduce the water supply to farm areas, flood valuable land, and damage fishing areas.

Industrial wastes and pollutants in rivers and lakes reduce water quality both for farming and recreation. These indirect costs due to ''lost ecological capital'' are estimated in energy units in the analysis carried out by Odum.[3,5,6] The 1974 IFIAS Workshop,[1] however, made no specific recommendations on a method or convention for analysis of these environmental costs. Lacking a proper framework and data, most energy analysts have not included this factor in their estimates, considering it outside the boundary being considered.

Some energy analysts have opted for the techniques used by Odum[5] as they include a larger range of factors and have a wider boundary than those defined at the 1974 IFIAS Workshop.[1] However, most analysts in the construction industry have adopted the IFIAS conventions as they are simpler and the calculations are more easily carried out. To gauge the extent of the overall amount of energy consumed for construction, we shall now look at the construction industries of two countries.

III. ENERGY USE IN THE CONSTRUCTION INDUSTRY

In this section the energy consumed for construction is compared with the amount going to other industries. A breakdown of the energy consumed by various sectors of the construction industry is also presented. The energy requirements of the construction industries of two countries, the U.S. and New Zealand, are reviewed.

In modern industrialized countries, the construction industry is largely factory based, with much of the building process occurring away from the building site. Site work consists mainly of component assembly. This situation is reflected in the low proportions of direct energy required by the construction industries of the U.K.,[7] the U.S.,[8] and New Zealand,[9] as illustrated in Figure 1.

The following brief overview of the energy requirements of the construction industries of two of these countries, the U.S. and New Zealand, will further quantify the energy required for the various stages of construction and for the different sectors of the construction industry. In many respects the two countries are markedly different. Their respective economies are giant and miniscule by world standards, and their degree of industrialization is quite different. Despite these differences, a breakdown of the energy requirements of the various sectors of the U.S. and New Zealand construction industries shows the proportions going to each sector to be very similar.[8,9]

A. The U.S. Construction Industry of 1967

The energy requirements of the U.S. construction industry in 1967 were studied in detail by Stein et. al.[4,8,10,11] Their findings were as follows: ''The energy required by the U.S.A.

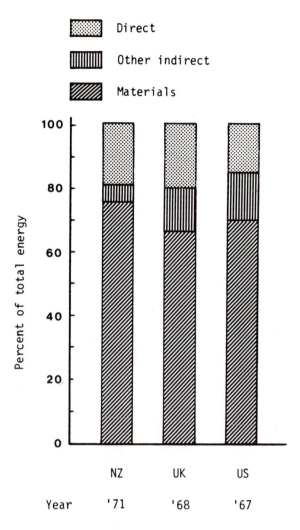

FIGURE 1.   Comparison of the breakdown of energy requirements
of the New Zealand, U.K., and U.S. construction industries.

construction industry represents 11.14 per cent of the total energy consumed in the United
States. This is equivalent to about 7900 PJ (7,500 × 10¹² Btu). New building construction
accounts for about 5.19 per cent of the total U.S.A. energy requirement while the remaining
5.95 per cent is distributed among building maintenance and other non-building construction
such as roads, railways, bridges, dams and others.''

Table 3 gives a detailed breakdown of the energy requirements for the various types of
new buildings. Almost half the energy requirements for new buildings went to residential
construction and the remainder to nonresidential buildings. Approximately 70% of the energy
requirements of the U.S. building industry were for the production and processing of building
materials. The ten largest material supply sectors contributed about half of this indirect
energy. Table 4 gives the energy inputs from the ten ''largest energy input sectors'' to new
building construction as a whole.

In terms of intensity of energy required per unit floor area for constructing new buildings,
hospitals and office buildings required the most, and residential buildings and warehouses
required the least energy, as illustrated in Table 5. The table also shows the proportion of
direct to indirect energy required for the various building types. For all building types, the

Table 3
U.S. BUILDING ENERGY USE (NEW CONSTRUCTION — 1967)

Construction	PJ	10^{12} Btu	Percentage
Residential — new	1195	1133	32.6
Residential — alteration	279	264	7.6
Hotels and motels	74	70	2.0
Hospitals	125	119	3.4
Factories	495	469	13.5
Commercial	586	556	16.0
Schools	467	443	12.7
Others	444	421	12.1
Total	3665	3475	100.0

From Stein, R. G., et al., Handbook of Energy Use for Building Construction, DOE/CS/20220-1, U.S. Department of Energy, 1980.

Table 4
INPUTS TO NEW BUILDING CONSTRUCTION FROM THE TEN LARGEST ENERGY INPUT SECTORS (U.S.)

Sector no.	Name	PJ	10^{12} Btu	Percentage of new building construction
3	Refined petroleum products	579	549	15.79
135	Sawmills	91	86	2.47
184	Asphalt	91	86	2.47
196	Bricks	111	105	3.04
206	Ready-mix concrete	318	301	8.69
240	Fabricated structural steel	206	195	5.62
241	Metal doors	107	101	2.98
243	Sheet metal work	101	96	2.75
245	Misc. metal work	100	95	2.72
373	Retail trade	99	94	2.71
	Others	1863	1766	50.76
	Total	3666	3474	100.00

From Stein, R. G. et al., Handbook of Energy Use for Building Construction, DOE/CS/20220-1, U.S. Department of Energy, 1980.

indirect energy makes up the major portion of the total energy requirement. A more detailed study by Kegel[12] of the energy requirements for the materials and construction of a community college showed that travel to the job site and temporary heating at the site make up more than half of the direct energy requirements. Table 6 outlines the results from Kegel's investigation.

B. The New Zealand Construction Industry of 1971—1972

The New Zealand construction industry accounted for 39 PJ (37×10^{12} Btu), roughly 10% of the total New Zealand energy requirement, in 1971—1972. That total includes new building construction, building maintenance, and other nonbuilding activities such as road, bridge, and aerodrome construction. The construction industry energy requirements are almost equally split between the building industry and other nonbuilding activities. The detailed breakdown of the building industry energy use in 1971—1972, as given in Table

Table 5
INTENSITY OF ENERGY REQUIREMENTS FOR BUILDING
CONSTRUCTION (U.S.)

Energy requirements per unit floor area

	Total		Direct		
Construction	GJ m^{-2}	10^3 Btu ft^{-2}	GJ m^{-2}	10^3 Btu ft^{-2}	Direct/ total (%)
Residential family	8.0	704	1.0	88	13
Hotels and motels	12.8	1126	2.9	255	23
Hospitals	19.6	1725	4.0	352	20
Industrial buildings	11.0	968	1.1	97	10
Office buildings	18.6	1637	4.0	352	22
Stores and restaurants	10.7	942	2.5	220	23
Religious buildings	14.3	1258	2.9	255	20
Warehouses	6.3	554	0.9	79	14

From Stein, R. G. et al., Energy Use for Building Construction CAC, EDRA Contract EY-76-S-02-2791, University of Illinois, Urbana and R. G. Stein and Associates, New York, 1976.

Table 6
ENERGY REQUIREMENTS FOR CONSTRUCTION
OF A COMMUNITY COLLEGE

GER per unit floor area

Materials	GJ m^{-2}	10^3 Btu ft^{-2}	Percentage
Steel	3.7	325.6	69
Concrete	0.7	61.6	12
Others	1.0	88.0	19
Total indirect	5.4	475.2	100
Activity			
Clearing site	0.2	17.6	9
Helicopter	0.12	10.6	5
Labor travel to job	0.46	40.5	20
Material delivery	0.20	17.6	9
Off-site fabrication	0.26	22.9	11
Temporary electric power	0.12	10.6	5
Temporary heat	0.80	70.4	34
Other	0.17	15.0	7
Total direct	2.33	205.2	100
Total GER	7.73	680.4	

7, shows that about half of this goes towards residential construction. Although the quantity of energy consumed by the New Zealand construction industry is considerably smaller than that of the U.S., the proportions going to the various sectors are very similar for the two countries.

However, the embodied energy per unit floor area, for constructing new buildings, is higher in the U.S. than in New Zealand. For New Zealand residential and commercial construction, the intensities are about 3300 and 4000 MJ m^{-2} (290,000 and 350,000 Btu

Table 7
NEW ZEALAND BUILDING INDUSTRY
ENERGY USE (NEW CONSTRUCTION
1971—1972)

Construction	PJ	10^{12} Btu	Percentage
Residential — new	7.8	7.4	44
Residential — alteration	1.3	1.2	7
Hotels	0.7	0.7	4
Hospitals	0.7	0.7	4
Factories	1.7	1.6	10
Commercial	2.4	2.3	13
Schools	1.3	1.2	7
Others	1.9	1.8	11
Total	17.9	16.9	100

Table 8
INPUTS TO BUILDING CONSTRUCTION FROM THE TEN
LARGEST ENERGY INPUT SECTORS (NEW ZEALAND)

NO.	Industry	TJ	10^9 Btu	Percentage of building industry
43	Sawmills	526	499	3.06
44	Planing and timb. pres.	661	627	3.84
46	Joinery	658	624	3.83
47	Pulp and paper	452	429	2.63
64	Bitumen	362	343	2.10
72	Cement	688	652	4.00
74	Conc products	521	494	3.03
75	Ready-mix conc	721	684	4.19
76	Nonmetallic	479	454	2.78
81	Metal products NEC	527	500	3.06
	Others	11605	11002	67.47
	Total	17200	16308	100.00

ft^{-2}) compared with figures of 8000 and around 13000 MJ m^{-2} (700,000 and 1,150,000 Btu ft^{-2}) in the U.S.

Construction energy requirements account for about 10% of the total energy consumption of both the U.S. and New Zealand. Energy required for the manufacture of materials makes up the largest proportion, amounting to some 70% of the total construction industry energy requirements. Thus the energy embodied in building materials is by far the most important component in the capital energy of a building. Of this 70% the "ten largest energy input industries" contribute approximately 30%. Table 8 lists the ten largest New Zealand energy input sectors. As in the U.S. building industry (Table 4), sawmills, cement, ready-mix concrete, and metal industries appear to be among the most energy intensive.

We shall consider the energy embodied in some common building materials in the next section.

IV. ENERGY COEFFICIENTS OF BUILDING MATERIALS

The energy coefficient of a material is the energy required to produce a unit of the material, e.g., the energy per unit mass or per unit volume. Values for some common building materials will be presented in this section. The process involved in the production of the building materials will also be outlined briefly to give an overview of the range of energy coefficients of different varieties of the same material and for the different degrees of refinement.

The energy coefficient of a material or process may not be the same for different countries and different years. This is because of variations in:

- Production sites
- Raw material quality
- Climate
- Processing methods
- Age and type of plant
- Efficiency of energy and material use, and
- Basic load requirements

There may also be fairly large variations in energy coefficients if the analyses are taken to different levels, that is, different system boundaries are drawn. The proportions of coal, oil, gas, and electricity making up the total energy requirements for each material will differ between countries, and this may further contribute to variations in the energy coefficient. Finally, different conventions and methods of evaluation will contribute to variations in the calculated results.

A. Transport Energy

Transport energy requirements often make up a high proportion of the total costs in the building industry because of the high weight to value ratio of building materials. These energy requirements average out at about 4.5 MJ t^{-1} km^{-1} (6900 Btu t^{-1} mile^{-1}) for road transport and 0.6 MJ t^{-1} km^{-1} (900 Btu t^{-1} mile^{-1}) for rail transport in the U.S. and New Zealand. In the case of air transport, the New Zealand figure is around 50% higher, at 15 MJ t^{-1} km^{-1} (22,900 Btu t^{-1} mile^{-1}) than that of the U.S.[13]

B. Quarrying Energy

In both the U.S. and New Zealand, a large proportion (about 90% by mass and cost) of quarry products go either directly or indirectly to the construction industry. The energy required for extracting natural sand and natural aggregates is about 17 and 19 MJ t^{-1}, respectively (16,100 and 18,000 Btu t^{-1}),[7,14] while that for extracting, crushing, and grinding crushed aggregate is about 80 to 90 MJ t^{-1} (76,000 to 85,000 Btu t^{-1}).[7,14,15]

C. Wood and Paper Products

The manufacture of wood and paper products starts off with relatively low energy intensity processes, logging, transport, and debarking. In the case of timber production, sawmilling and treatment are also involved, the final treatment process being the most energy intensive.

For sheet timber products such as fiberboards, paperboards, particle boards, veneers, and plywood, pulping, pulp drying, and board making are involved. Again, the final step is the most energy intensive and makes up about half the process energy requirement.

Paper production involves pulping, drying, and papermaking. These three steps consume more than 90% of the gross energy requirement, which can amount to figures in the region of 40 to 60 MJ kg^{-1} (17,200 to 25,750 Btu lb^{-1}).

D. Cement, Concrete, and Plaster Products

1.Cement

The manufacture of portland cement consists of three major stages: raw material processing (quarrying, crushing, drying, and grinding limestone and marl), pyroprocessing, and finish grinding. Pyroprocessing takes up about 80% of the process energy requirements. Energy requirements of around 9 MJ kg^{-1} (3900 Btu lb^{-1}) have been suggested,[9] but the amount can vary significantly with the manufacturing process.

2. Concrete

For concrete, the energy requirement varies widely depending on the cement, aggregate, sand ratio, and transport requirements. It ranges from about 1.6 to 2 MJ kg^{-1} (690 to 860 Btu lb^{-1}) with a stronger concrete usually having a higher gross energy coefficient.

3. Plaster

Plaster is made from crushing and dehydrating gypsum. One major building product that uses plaster is gypsum board. It is made by mixing plaster with various fillers to produce a paste which is then formed between sheets of paper board and dried. The energy coefficient for gypsum board is about 60 to 80 MJ m^{-2} (5300 to 7000 Btu ft^{-2}).[8,16]

E. Glass

The main steps in the manufacturing process are

- Mixing of silica, limestone, sodium carbonate, and waste glass
- Melting of raw materials
- Refining the molten glass
- Forming the refined glass
- Finishing

The melting and refining steps consume about 80% of the energy requirement, which can range from 8 to 30 MJ kg^{-1} approximately (3400 to 12,900 Btu lb^{-1}).

F. Clay Products

The production of structural clay can be divided into:

- Crushing, milling, blending, moistening, and de-aerating the clay
- Extruding or forming in a mold
- Drying to remove moisture
- Firing in a kiln, which is the most energy-intensive stage

Average gross energy requirement figures of around 7 MJ kg^{-1} (3000 Btu lb^{-1}) have been suggested.[9]

G. Paints

Paints have the following constituents: binders (or medium), driers and catalysts (hardeners), pigments, extenders, solvents, thinners, emulsifiers, flattening agents, and gelling agents. Depending on the type of paint and thus the chemical constituents, the energy requirements for manufacture vary considerably, from about 100 to 200 MJ ℓ^{-1} (360,000 to 720,000 Btu gal^{-1}).

H. Plastics

The manufacture of plastics can be divided into four major steps:

- Acquisition of feedstocks
- Manufacture of monomeric and other inputs to polymerization
- Polymerization
- Product fabrication

The monomer production and polymerization processes are the most energy- and capital-intensive steps, amounting to between 80 and 90% of the total process energy requirement

Table 9
ENERGY COEFFICIENTS OF THE MAIN BUILDING
MATERIALS FOR THE U.S., THE U.K., AND NEW
ZEALAND (IN SI UNITS)

Product	Units	GER (MJ/unit)		
		U.S.	U.K.	N.Z.
Finished timber	m³	3000	4480	4692
	kg	4.3	6.4	6.7
Paper	kg	61	42	26*
Cement	kg	9.4	5.9	9
Concrete	kg	1	1.7	2
Plaster	kg	5.9	3.2	6.7*
Glass	kg	8.4	22	31.5
Structural clay	kg	7.2	2	6.9
Paints	liter	120	115	—
Plastics general	kg	—	159	—
Polyethylene	kg	163*	104*	112*
PVC	kg	87*	85*	96*
Polypropylene	kg	157*	171*	—
Polystyrene	kg	140*	96*	—
Iron	kg	16	23.8	16*
Steel	kg	35—100	37—47	35—50
Copper	kg	106—166	45	—
Aluminum	kg	211—265	96	130—154*
Lead	kg	26	25	—
Zinc	kg	38	67	—

Note: Energy coefficients given are to IFIAS Level 4 except those marked
* which are to Level 2. Table is summary of data from Baird and
Chan.[9]

for the finished product. The energy requirement also varies widely for the different types
of plastics and for different countries.

I. Metal Products

The production of most metals can be considered in four main stages as follows:

- Mining and quarrying of the ore, coke, limestone, and other raw materials
- Reduction of the ore in a furnace, usually the most energy-intensive stage
- Refining and alloying
- Casting, fabricating, and machining into the finished products

The energy requirements are affected to a large extent by the proportion of recycled scrap
metal to raw ore, the latter requiring much more energy. In the U.K., for example, for steel
with 50% scrap, the energy requirement is 47.5 MJ kg^{-1} (20,400 Btu lb^{-1}) but with 100%
scrap, it is 23.4 MJ kg^{-1} (10,000 Btu lb^{-1}).

Iron and steel are the most extensively used metals in the construction industry. Depending
on product and finish, the refining and fabrication stages could be very energy intensive.
A semifinished slab of steel, for example, may have an energy coefficient as low as 28 MJ
kg^{-1} (12,000 Btu lb^{-1}), whereas a thin sheet of special heat treated alloy steel may have
an energy coefficient as high as 210 MJ kg^{-1} (90,100 Btu lb^{-1}). Figures for a range of
metal products are given in Tables 9 and 10.

Table 10
ENERGY COEFFICIENTS OF THE MAIN BUILDING MATERIALS FOR THE U.S., THE U.K., AND NEW ZEALAND (IN IP UNITS)

Product	Units	GER (Btu/unit) U.S.	GER (Btu/unit) U.K.	GER (Btu/unit) N.Z.
Finished timber	ft³	80,500	120,200	125,900
	lb	1,845	2,747	2,876
Paper	lb	26,180	18,026	11,159*
Cement	lb	4,034	2,532	3,863
Concrete	lb	429	730	858
Plaster	lb	2,532	1,373	2,876*
Glass	lb	3,605	9,442	13,519
Structural clay	lb	3,090	858	2,961
Paints	gal (U.S.)	431,000	413,000	—
Plastics general	lb	—	68,240	—
Polyethylene	lb	69,956*	44,635*	48,068*
PVC	lb	37,339*	36,480*	41,201*
Polyproplene	lb	67,381*	73,390*	—
Polystyrene	lb	60,085*	41,201*	—
Iron	lb	6,867	10,214	6,867*
Steel	lb	15,021—42,981	15,880—20,171	15,021—21,459
Copper	lb	45,493—71,244	19,313	—
Aluminum	lb	90,557—113,733	41,201	55,793—66,094*
Lead	lb	11,159	10,730	—
Zinc	lb	16,309	28,755	—

Note: Energy coefficients given are to IFIAS Level 4 except those marked * which are to Level 2. Table is summary of data from Baird and Chan.[9]

J. Summary

Tables 9 and 10 summarize the energy coefficients of the main building materials for the U.S., the U.K., and New Zealand. Although there are differences in energy coefficients for the three countries which could be as much as 100%, the energy coefficients for the same materials are all of the same order of magnitude.

V. ENERGY REQUIREMENTS OF BUILDING COMPONENTS

The purpose of this section is to demonstrate the worth of the Energy Analysis method. To do so, the data from the review of the previous section of the energy coefficients of building materials is used to obtain the energy requirements of building components. The following comparative analysis of some typical building components will serve to illustrate the range of energy requirements for these various alternatives. Some common, domestic-scale, floor, wall, and roof assemblies have been selected.

A parallel money value comparison is given, as well as a cost-in-use analysis. The energy requirements and dollar values apply to New Zealand in December 1979, but the energy requirements should be similar for other countries for the same construction.

A. Floor Construction

The two main types of floor construction for houses are the suspended timber floor and the concrete slab on ground. The typical construction details of each are illustrated in Figures 2 and 3. Tables 11 and 12 present the estimation of the Gross Energy Requirement (GER) for each case. It will be seen that the suspended timber floor has a GER of 733 MJ m^{-2} (64,500 Btu ft^{-2}) while that for the concrete slab is estimated to be 1014 MJ m^{-2} (89,200 Btu ft^{-2}).

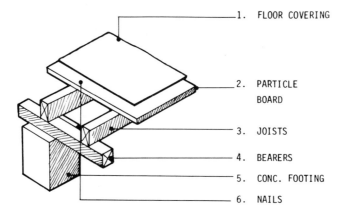

1. FLOOR COVERING

2. PARTICLE BOARD

3. JOISTS

4. BEARERS

5. CONC. FOOTING

6. NAILS

FIGURE 2.   Typical construction detail — suspended timber floor.

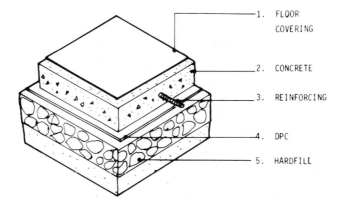

1. FLOOR COVERING

2. CONCRETE

3. REINFORCING

4. DPC

5. HARDFILL

FIGURE 3.   Typical construction detail — concrete slab on ground floor.

Table 11
ESTIMATE OF COSTS AND GER — SUSPENDED TIMBER FLOOR

I	Cost per unit floor area		GER per unit floor area	
	N.Z.$m^{-2}	N.Z.$ft^{-2}	MJ m^{-2}	Btu ft^{-2}
Floor covering — vinyl	12.50	1.16	380	33,440
Particle board, 20 mm(0.75 in.) high density	13.86	1.29	200	17,600
Joists, 150 × 50 mm(6 × 2 in.) at 450 mm (18 in.) crs	9.17	0.85	72	6,336
Bearers, 100 × 100 mm (4 × 4 in.)	2.09	0.19	22	1,936
Concrete footings	3.08	0.29	53	4,664
Nails	—	—	6	528
Totals	40.70	3.78	733	64,504

The latter figure is expected to drop when New Zealand (in common with many other countries) completes the conversion of its cement manufacturing from the wet to the less energy-intensive dry process. Hence, it is anticipated that there will be little significant difference between these two methods of floor construction, in terms of GER, in the case of future domestic buildings.

Table 12
ESTIMATE OF COSTS AND GER — CONCRETE SLAB ON GROUND FLOOR

Item	Cost per unit floor area		GER per unit floor area	
	N.Z.$m^{-2}	N.Z.$ft^{-2}	MJ m^{-2}	Btu ft^{-2}
Floor covering — vinyl	12.50	1.16	380	33,440
Concrete floor slab, 17.5 MPa, 100 mm (4 in.)	9.40	0.87	400	35,200
Reinforcement	2.21	0.21	70	6,160
Damp proof course, 0.25 mm (0.01 in.) — polythene sheet	1.93	0.18	60	5,280
Hardfill, 100 mm (4 in.)	2.80	0.26	104	9,152
Totals	28.84	2.68	1,014	89,232

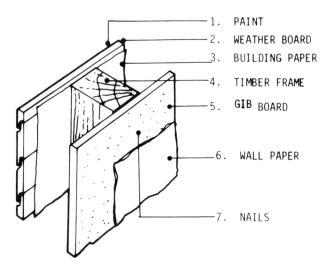

1. PAINT
2. WEATHER BOARD
3. BUILDING PAPER
4. TIMBER FRAME
5. GIB BOARD
6. WALL PAPER
7. NAILS

FIGURE 4.   Typical construction detail — timber framed wall with weatherboard.

B. Wall Construction

Three common domestic-scale exterior wall systems for timber frame construction are compared in Figures 4 to 6. All three walls have the same thermal resistance, without insulation, of 0.5 m^2 °C W^{-1} (2.8 ft^2 hr °F Btu^{-1}).

As will be seen from a study of Tables 13 to 15, the brick veneer wall has a GER over six times greater than the weatherboard wall. This is a reflection of the relatively larger energy requirement for firing bricks leading to a larger GER. The GER of the concrete block wall lies midway between the other two constructions.

For a standard single story timber framed house (see later for a full description), the wall area will be about 93 m^2 (1000 ft^2). The extra energy required for this house, if brick instead of weatherboard were used, would be about 90 GJ (85 × 10^6 Btu). In energy terms it would "cost" about 30% more. The labor requirements and transport energy requirements for the brick wall system will also be higher.

The energy required for maintenance of the weatherboard walls in the form of painting, say, every 3 years will be greater than for the brick wall. However, even when this maintenance energy is added, the total embodied energy for the weatherboard wall system will still be less than that for the brick wall system after 50 years.

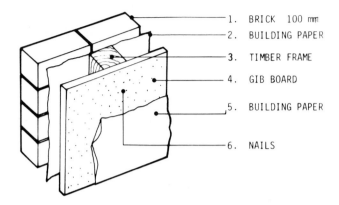

FIGURE 5. Typical construction detail — timber framed wall with brick veneer.

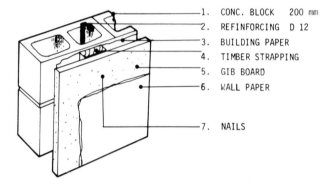

FIGURE 6. Typical construction detail — concrete block wall.

Table 13
ESTIMATE AND COSTS OF GER — TIMBER FRAMED WALL WITH WEATHERBOARD CLADDING

Item	Cost per unit wall area		GER per unit wall area	
	N.Z.\$m^{-2}	N.Z.\$ft^{-2}	MJ m^{-2}	Btu ft^{-2}
Paint — primer and 2 coats	3.50	0.33	43	3,784
Weatherboards, ex. 150 × 25 mm Hardwood (6 × 1 in.)	24.45	2.27	18	1,584
Building paper	0.63	0.06	6	528
Timber frame, 100 mm (4 in.) studs 450 mm (18 in.) crs, ex. 100 × 50 (4 × 2 in.) dwangs 800 mm (30 in.) crs, ex. 100 × 50 (4 × 2 in.)	8.26	0.77	64	5,632
Gypsum board 9.5 mm (0.4 in.)	5.85	0.54	50	4,400
Wallpaper	2.59	0.24	12	1,056
Nails	—	—	5	440
Totals	45.28	4.21	198	17,424

Table 14
ESTIMATE OF COSTS AND GER — TIMBER FRAMED WALL WITH BRICK VENEER CLADDING

	Cost per unit wall area		GER per unit wall area	
Item	N.Z.\$m^{-2}	N.Z.\$ft^{-2}	MJ m^{-2}	Btu ft^{-2}
Brick, 100 mm (4 in.) — plain red	35.50	3.30	1,150	101,200
Building paper	0.63	0.06	6	528
Timber frame, 100 mm (4 in.)	8.26	0.77	64	5,632
Studs 450 mm (18 in.) crs,				
ex. 100 × 50 (4 × 2 in.)				
Dwangs 800 mm (30 in.) crs,				
ex. 100 × 50 (4.2 in.)				
Gypsum board 9.5 mm (0.4 in.)	5.58	0.54	50	4,400
Wallpaper	0.59	0.05	12	1,056
Nails	—	—	2	176
Totals	50.83	4.72	1,284	112,992

Table 15
ESTIMATE OF COSTS AND GER — CONCRETE BLOCK WALL

	Cost per unit wall area		GER per unit wall area	
Item	N.Z.\$m^{-2}	N.Z.\$ft^{-2}	MJ m^{-2}	Btu ft^{-2}
Concrete blocks type 2001	26.18	2.43	404	35,552
Reinforcing D12, 600 × 800 mm (24 × 30 in.)	3.37	0.31	272	23,936
Building paper	0.63	0.06	6	528
Timber strapping, 600 mm (24 in.) crs	2.87	0.27	6	528
Gypsum board, 9.5 mm (0.4 in.)	5.85	0.54	50	4,400
Wallpaper	2.59	0.24	12	1,056
Nails	—	—	5	440
Totals	41.49	3.85	755	66,440

C. Roof Construction

Galvanized iron and concrete tile (see Figures 7 and 8) are the two most common roofing materials for domestic-scale buildings in New Zealand. Comparison of these two roofing systems (see Tables 16 and 17) indicates that the galvanized iron roof has a higher GER, mostly due to the higher GER for the galvanized iron. This conclusion, however, applies only to domestic-scale buildings and may not apply to larger-scale buildings where long-run galvanized iron roofing can be laid almost flat and requires less support than tile roofs, which need a minimum slope of about 17°, and therefore involve more trusses, purlins, and a greater roof area. The differences in these parameters, for the case of domestic-scale construction, are small, and the roof cladding GER appears to be the more important factor. However, the other parameters may become significant in larger-scale roofs.

Galvanized iron roof systems also have a higher maintenance cost, requiring painting every 5 to 10 years. The cost-in-use comparison of concrete tile and galvanized iron roofs, presented in Table 18, indicates that the latter is almost twice as expensive.

As can be seen from this brief review, there can be a considerable range in energy requirements per unit component area, even for some fairly basic and relatively similar constructions. Table 19 provides a comparative summary of all the building assemblies considered.

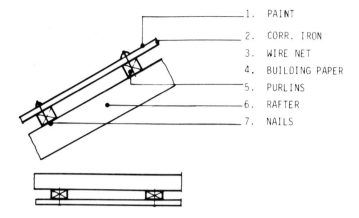

1. PAINT
2. CORR. IRON
3. WIRE NET
4. BUILDING PAPER
5. PURLINS
6. RAFTER
7. NAILS

FIGURE 7. Typical construction detail — galvanized iron roof.

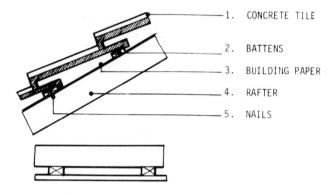

1. CONCRETE TILE
2. BATTENS
3. BUILDING PAPER
4. RAFTER
5. NAILS

FIGURE 8. Typical construction detail — concrete tile roof.

Table 16
ESTIMATE OF COSTS AND GER — GALVANIZED IRON ROOF

Item	Cost per unit roof area		GER per unit roof area	
	N.Z.$m^{-2}	N.Z.$ft^{-2}	MJ m^{-2}	Btu ft^{-2}
Paint, primer and 2 coats	2.55	0.24	31	2,728
Corrugated iron, 20-gauge	7.75	0.72	360	31,680
Wire mesh 50 × 1 mm (2 × 0.04 in.)	1.18	0.11	40	3,520
Building paper	0.63	0.06	6	528
Purlins 75 × 50 mm (3 × 2 in.), 600 mm (24 in.) crs	3.24	0.30	26	2,288
Rafter 200 × 50 mm (8 × 2 in.), 900 mm (36 in.) crs	3.34	0.31	38	3,344
Nails	—	—	7	616
Totals	18.69	1.74	508	44,704

VI. ENERGY REQUIREMENTS OF WHOLE BUILDINGS

The integration of the energy requirements of the appropriate building assemblies considered in the previous section, to derive the energy requirements of whole buildings, will now be demonstrated. A sample capital energy requirement estimate for a single story timber framed house is introduced. Using this as an index of reference, the effects of variation in

Table 17
ESTIMATE OF COSTS AND GER — CONCRETE TILE ROOF

Item	Cost per unit roof area		GER per unit roof area	
	N.Z.\$m^{-2}	N.Z.\$ft^{-2}	MJ m^{-2}	Btu ft^{-2}
Concrete tiles	12.85	1.19	100	8,800
Battens 25 × 50 mm (1 × 2 in.), 100-gauge	—	—	19	1,672
Building paper	0.63	0.06	6	528
Rafter 200 × 50 mm (4 × 2 in.), 750 mm (30 in.) crs	4.32	0.40	49	4,312
Nails	—	—	2	176
Totals	17.8	1.65	176	15,488

Table 18
COMPARISON OF COST IN USE OF CONCRETE TILE AND GALVANIZED IRON ROOFS (NEW ZEALAND — 1979)

Material	Concrete tile (\$)	Galvanized iron (\$)
Service life (years)	50	50
Gross capital cost	2269.50	2383.00
Annual equivalent capital cost	124.00	129.40
Maintenance	—	108.40
Cost-in-use	124.00	237.80

Table 19
COMPARISON OF COSTS AND GER OF SOME COMMON, DOMESTIC-SCALE FLOOR, WALL, AND ROOF ASSEMBLIES

Floor, wall, or roof assembly	Cost per unit area		GER per unit area	
	N.Z.\$m^{-2}	N.Z.\$ft^{-2}	MJ m^{-2}	Btu ft^{-2}
Floors				
Suspended timber floor	40.70	3.78	733	64,504
Concrete slab on ground	28.84	2.68	1014	89,232
Walls				
Timber framed wall with weatherboard cladding	45.28	4.21	198	17,424
Timber framed wall with brick veneer cladding	52.83	4.91	1284	112,992
Concrete block wall	41.49	3.85	75	66,440
Roofs				
Galvanized iron roof	18.69	1.74	508	44,704
Concrete tile roof	17.80	1.65	176	15,488

the building materials used, on the capital energy requirement, are estimated. A comparison is also made with alternative construction systems.

The standard house introduced here, although typical of houses in New Zealand, may not be typical of those in the U.S. or U.K. It serves only as an illustration of the principles and methods involved.

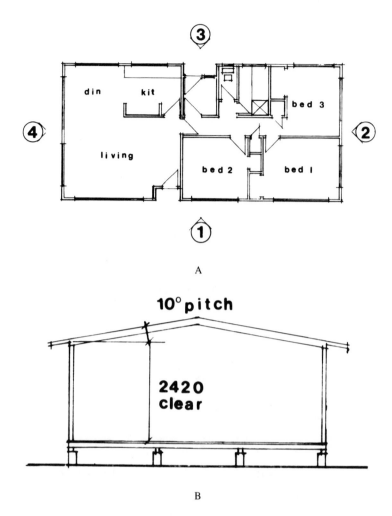

FIGURE 9. BIAC standard house — scale 1:100. (A) plan; (B) X-section. (C) elevations.

A. Energy Requirements of a Standard House

The building selected is the New Zealand Building Industry Advisory Council (BIAC) standard house, which is fairly typical of a large number of houses in New Zealand. A brief specification for the house is as follows: 94 m² (1012 ft²) floor area, three bedrooms, level site, precast concrete piles, concrete steps, suspended timber floor, weatherboard wall cladding, corrugated iron clad gable roof, timber joinery, particle board floor, gypsum board interior wall lining, sloping ceiling with exposed rafters to dining room and lounge, flat ceiling to other areas, separate shower, bath and laundry, separate W.C., 12 electric lights and 16 power points. Plans, elevations, and sections of the house are illustrated in Figure 9.

As an example of the computation involved, a detailed list of materials, quantities, costs, and energy requirements for the floor construction of the standard house is given in Table 20. Table 21 summarizes the data for the major components of the house as a whole. The estimates are GERs up to Level 4 as defined by the 1974 IFIAS Workshop.[1] The GER values include material transport to the site, on-site energy requirements, and material wastes. Parallel 1979 New Zealand dollar value estimates are also given. The prices represent average installed costs in New Zealand dollars in December 1979 and are based on the ruling rate

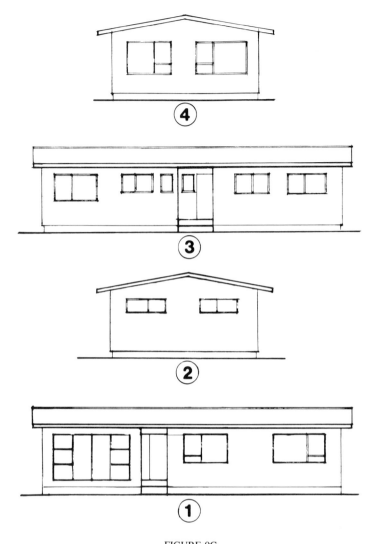

FIGURE 9C

of material costs along with proper allowances, contractor's on-site and overhead charges, etc.

B. Comparison with Alternative Constructions

As shown earlier, the use of different materials and construction techniques results in differences in intensity of energy requirement. Moreover, even for the same building type, the intensity of energy requirement shows a considerable range. This is illustrated in Table 22 where it will be seen that domestic buildings in the U.K. and the U.S. appear to be more energy intensive than those in Australia and New Zealand. These variations are most likely due to the different building practices and material GERs in these four countries. However, the methods of calculation and conventions adopted in these countries may also contribute to the variations.

C. Comparison of Construction and Operating Energy Requirements

The ratio of the annual operating energy requirements to the construction GER could be important when comparing alternative designs. This ratio provides an index by which dif-

Table 20
ILLUSTRATION OF THE METHOD OF ESTIMATING THE ENERGY REQUIREMENTS OF THE FLOOR OF THE NEW ZEALAND BIAC STANDARD HOUSE

Floor	Unit	Quantity	Cost 1979 (N.Z.$)	Rate (MJ/unit)	Waste %	Amount MJ	Amount 10³ Btu
Excavation and ground work	m³	1	17	100	0	100	95
Precast concrete piles and	No	36	290	—	—		
footings 600 × 200 × 200	m³	1.18		4,780	10	6,200	5,878
on 300 × 300 × 100							
Bearers	m	56	197				
	m³	0.56	—	4,690	10	2,890	2,740
Floor joists 150 × 50 mm	m	245	862				
at 450 mm crs	m³	1.84	—	4,690	10	9,500	9,006
Particle board flooring	m²	95	1,317				
High density 20 mm	m³	1.9	—	12,890	20	29,390	27,862
Nails and fasteners	kg	12	—	35	5	440	417
Concrete steps	m³	3	282	3,840	10	12,670	12,011
Reinforced steel	kg	20	17	35	10	770	730
Total floor			2,982			61,960	58,739

Table 21
BIAC STANDARD HOUSE — SUMMARY OF COSTS AND GER OF COMPONENTS

Components	Cost 1979 ($N.Z.)	%	GER GJ	GER 10⁶Btu	GER %
Floor	2,980	12.2	62	59	17.4
Walls	7,820	32.0	50	47	14.0
Roof	2,160	8.8	66	63	18.5
Joinery	3,320	13.5	62	59	17.4
Plumbing	3,540	14.4	28	27	7.8
Electrical	1,150	4.7	5	5	1.4
Finishes	2,850	11.6	57	54	16.0
Insulation	690	2.8	16	15	4.5
Preliminaries	—	—	11	10	3.0
Total	24,510	100	357	339	100
per m² floor area	261		3.8	—	
Per ft² floor area	24		—	0.33	

Table 22
COMPARISON OF GER FOR HOUSE CONSTRUCTION IN FOUR COUNTRIES

House	Floor area m²	Floor area ft²	GER GJ m⁻²	GER 10³ Btu ft⁻²	Ref.
U.K., three bedroom semidetached	100	1076	4.0	352	17
U.K., standard midterrace house	96	1033	7.0	616	18
U.S., single-family	—	—	8.0	704	8
Aust. one story brick veneer slab on ground	144	1550	3.6	317	19
N.Z., timber frame house BIAC	94	1012	3.8	334	9
N.Z., timber frame house NZIV model house	93	1001	3.6	317	9

Table 23
RATIO OF CONSTRUCTION TO
OPERATING ENERGY REQUIREMENTS
FOR THE HOUSE CONSTRUCTIONS IN
FOUR COUNTRIES

House	Construction/operating (years)	Ref.
U.K., three bedroom semidetached	4.8	17
U.K., standard mid-terrace house	3.1	18
U.S., one family	6	8
Australia one story brick veneer slab on ground	6	19
N.Z., timber frame house BIAC	9.8	9
N.Z., timber frame house NZIV model house	9	9

fering buildings may be judged and alternative investments in energy conservation compared. For nondomestic buildings, the ratio will be important for examining the energy cost effectiveness of mechanical cooling and heating systems. The ratio can also be used to gauge the energy effectiveness of thermal insulating systems.

Table 23 outlines the typical construction to operating energy requirement ratios for the houses listed in Table 22. This data is consistent with the assumption that the value of the ratio will be lower in a location with a more severe climate, if other factors are equal, because of the higher operating energy requirement. Possibly most notable of all is the relatively short time (between 3 and 10 years) it takes for an amount of energy, equivalent to the initial capital energy requirement, to be consumed in the houses.

D. Energy Requirements of Variations to the Standard House

To examine the effect of differences in materials on the embodied energy total for construction, the total GER of the New Zealand BIAC standard house was estimated for a range of common materials. The results are presented in Table 24.

The lowest energy option is the house with weatherboard wall cladding, suspended timber floor, and tile roofing, while the highest energy option has brick veneer walls with a concrete slab on ground floor and galvanized iron roofing. The difference is about 2100 MJ (2.0×10^6 Btu). Thus the GER for construction of a house can be about 60% more, simply due to differences in commonly available construction materials. This is equivalent to about 4 years annual operating energy consumption. The selection of basic building materials for mass housing can thus have significant energy implications.

E. Summary

The GER for construction of houses varies from about 3 to 8 MJ m^{-2} (260,000 to 700,000 Btu ft^{-2}) of floor area, depending on the type and materials of construction. A house with basically the same structure, for example, can require up to 60% more capital energy, as a result of the designer's choice of building materials. The capital energy requirement for housing in the U.S. and the U.K. appears to be higher than in Australia and New Zealand due to differences in building practices and materials.

The ratio of construction to operating energy requirement can range from about 3 to 10

Table 24
VARIATION IN TOTAL GER FOR THE STANDARD
HOUSE DUE TO DIFFERENCES IN CONSTRUCTION
-MATERIALS

	GJ m^{-2}		10^3 Btu ft^{-2}	
	Roofing material		Roofing material	
Wall	Galvanized iron	Concrete tile	Galvanized iron	Concrete tile
Concrete Floor				
Weatherboard	4.4	3.9	387	343
Concrete block	4.7	4.3	414	378
Brick veneer	5.5	5.1	484	449
Timber Floor				
Weatherboard	3.8	3.4	334	299
Concrete block	4.2	3.7	370	326
Brick veneer	4.9	4.5	431	396

years. This ratio is important for gauging the energy cost effectiveness of "energy saving" construction and operating systems.

There are various other applications for Energy Analysis and these are described in the next section.

VII. DESIGN APPLICATIONS AND OPPORTUNITIES

In this final section of the chapter, the various applications of Energy Analysis available to building designers are outlined. The integrating of capital energy cost analysis into current design practices and procedures is also discussed.

A. Design Applications

For the designer, the most obvious and common applications of Energy Analysis are for:

- Energy cost-benefit evaluation of alternative new design options
- Energy cost-benefit analysis of renovation as against new construction

Energy Analysis can also be applied at many different levels of building design evaluation, from the detailed to the general. Some examples are listed as follows:

Building Components
- Comparing the energy costs of various structural, construction and mechanical systems
- Identifying high energy cost components
- Evaluating the benefits of increased insulation or double glazing

Individual Buildings
- Comparing the capital energy costs, operating costs, maintenance costs, and cost-in-use of the various design options
- Comparing the capital energy requirements of particular designs with those of standard designs, national averages and other building types

Groups or Particular Types of Buildings
- Identifying the designs or types of constructions which require large amounts of particular fuels (e.g., imported oil)
- Determining the overall energy benefit of many of the recent "energy saving" designs such as passive solar heating, solar water heating, and solar cooling
- Evaluating the effects on overall energy costs of the use of alternative building materials, structural and construction systems for standard buildings, mass housing, and large-scale projects
- Determining the average energy requirements for particular building types and identifying the structural and construction systems that have high energy requirements

B. Methods of Calculation

In most cases, the calculation of the embodied energy for building construction involves the preparation of a bill of quantities of materials and the application of the appropriate energy coefficients. As the preparation of a dollar value bill of quantities is already carried out for most projects, it is a relatively straightforward procedure to calculate the capital energy requirement if the energy coefficients of building materials are available. However, the bill of quantities for even an average size project can be a rather large document and parallel energy calculations by manual methods can be very tedious. It is therefore unlikely that calculations will be carried out by manual methods, unless there is a legislative requirement to do so.

There is a strong move towards the use of computers for the preparation of bills of quantities, design evaluation, and even the drafting and preparation of working drawings. It is a relatively simple procedure to adapt existing programs that carry out these evaluation and design functions to calculate the energy requirements as a parallel exercise to the dollar value computations. The programs should be able to draw from a readily available data bank which need to be updated regularly. Such data banks are becoming increasingly available.[9]

C. Conclusions

With the introduction of many "low energy designs" and "energy saving options", it is essential to know if, over the life of a building, the savings in operating energy requirements exceed the added capital energy cost of these energy saving components. If it does, then the number of years of operation of the component needs to be known. These and various other vital questions are left unanswered because designers do not have the time, or the data base, to carry out Energy Analysis as part of their normal procedures. The recent wider use of computers for design evaluations, and the availability of embodied energy data in a readily useable form, should enable Energy Analysis to be incorporated into most stages of design.

In the next chapter, we move on to the second of our designer concerns, the prediction of the energy used during the operation of the building.

REFERENCES

1. I.F.I.A.S., Energy analysis workshop on methodology and convention, in Int. Fed. Inst. Adv. Study, Workshop Report No. 6, Stockholm, 1974.
2. I.F.I.A.S., Workshop on energy analysis and economics, in Int. Fed. Inst. Adv. Study, Workshop Report No. 9, Stockholm, 1975.
3. **Odum, H. T.,** *Energy Analysis, Energy Quality and Environment,* Energy Analysis: A New Public Policy Tool, Westview Press, 1978.
4. **Stein, R. G., et. al.,** Handbook of Energy Use for Building Construction, DOE/CS/20220-1, U.S. Department of Energy, Washington, D.C., 1980.
5. **Odum, H. T.,** *Environment Power and Society,* Interscience, New York, 1971.
6. **Odum, H. T.,** Energy, ecology and economics, *Ambio,* 2, 6, 220, 1973.
7. **Pick and Becker,** Direct and indirect uses of energy and materials in engineering and construction, *Applied Energy,* 1, Applied Science Publishers Ltd., 1975.
8. **Stein, R. G., et al.,** Energy Use for Building Construction EDRA Contract EY-76-S-02-2791, CAC, University of Illinois, Urbana and R. G. Stein and Associates, New York, 1976.
9. **Baird, G., and Chan, S. A.,** Energy Cost of Houses and Light Construction Buildings and Remodelling of Existing Houses, Report No. 76, New Zealand Energy Research and Development Committee, University of Auckland, Auckland, New Zealand, 1983.
10. **Stein, R. G. et. al.,** Energy Use for Building Construction — Supplement, CAC, University of Illinois, Urbana and R. G. Stein and Associates, New York, 1977.
11. **Stein, R. G.,** Energy cost of building construction, *Energ. Build.,* 1, 27, 1977.
12. **Kegel, R. A.,** The energy intensity of building materials, *Heat. Piping, Air Cond.,* p. 37, June 1975.
13. Beca, Carter, Hollings and Ferner, Energy in Transport, Vol. 1, Report No. 27, New Zealand Energy Research and Development Committee, University of Auckland, Auckland, New Zealand, 1977.
14. **Hawthorne, R. K.,** Energy Requirements for Road Pavements, Road Research Unit, Newsletter No. 55, 16, National Roads Board, Wellington, New Zealand.
15. **Haseltine, B. A.,** Comparison of energy requirements for building materials and structures, *Struct. Eng.,* 53, 9, 357, 1975.
16. **Chapman, P. F.,** The energy cost of materials, *Energ. Policy,* p. 47, March 1975.
17. **Brown and Stellon,** The material account, *Built Environ.,* p. 415, August 1974.
18. **Markus, T. and Slesser, M.,** Housing Energy Economics Pilot Study, OP 76111, Scottish Development Department Research Project, University of Strathclyde, Glasgow, 1976.
19. **Hill, R. K.,** Gross Energy Requirements of Building Materials, in Conf. on Energy Conservation in the Built Environment, Department of the Environment, Housing and Community Development, Sydney, March 1978.

Chapter 6

DESIGNER CONCERNS — SYSTEMS ENERGY CONSUMPTION

I. INTRODUCTION

The mechanical and electrical systems of a building consume energy to provide transportation, lighting, ventilation, heating, and cooling for the building occupants. These systems have an operating energy requirement that is, over the life of the building many times larger than that needed to construct the building, the capital energy requirement discussed in Chapter 5.

This chapter concentrates on how a building designer can incorporate concern for the operating energy requirement of a building (or systems energy consumption) at the design stage.

To quantify systems energy consumption, techniques for predicting the operating energy requirement of a building are usually necessary. In this chapter such techniques are examined to illustrate their common features, data requirements, limitations, and assumptions.

The results of studies of building energy use in practice are presented. These results are contrasted with the commonly predicted influences of building design features, and reasons for discrepancies are suggested. The importance of the occupants to the operating energy requirement of a building is highlighted, and their potentially positive role in controlling building systems is discussed. The chapter concludes by suggesting ways in which the feedback loop between prediction and practice can be improved.

II. BACKGROUND TO DESIGN CONCERNS

The concern for energy-conscious design must be placed in the context of the constraints on design choices. This section deals briefly with the background to these constraints.

A. Exogenous Constraints

Only rarely does a building designer have any influence on the factors discussed in Chapter 4. The external climate, the building size, and the activities housed are all exogenous factors, and hence constrain design choices. Compliance with local and national building codes, and limitations (see Chapter 3) on what is feasible in terms of both technology and cost, further limit these choices.

Certain design choices are largely dictated by exogenous factors. For example, if the available building site does not receive any direct winter solar radiation then the option of direct solar heating is not available to the building designer. The planned construction of nearby buildings may also reduce available solar options. An honest appraisal of the constraints on a building designer will show that a free choice in such factors as the building form, external fabric, and type of energy-consuming services is not usually available.

B. The Need to Predict Energy Use

Given the constraints (exogenous factors) on a particular building design there will be a whole spectrum of viable solutions from which the "best" design may be chosen. In present nonresidential construction the capital and operating energy costs of energy-consuming systems often comprise a significant fraction of the total building cost. In a climate of high energy costs, the operating energy consumption is a vital factor in choosing the "best" building design. There are often choices between cheaper capital cost solutions with higher anticipated operating costs and more expensive capital cost solutions with lower anticipated operating costs.

Determining the capital cost of conventional energy-consuming equipment is possible within the constraints of inflation and normal business uncertainty. Predicting the operating cost is a much more difficult problem as both energy use and future energy prices are unknown. Their "one-off" nature and the complexity of the thermal behavior of buildings causes difficulty in the prediction of their energy use. Experience gained from existing buildings must be used with caution as the building solution under consideration will usually differ in some respect.

Expectations of future energy prices are heavily influenced by current political and economic events such as a temporary oil surplus caused by an international recession. Since a building may be expected to consume energy for several decades, a considerable degree of judgment is needed to set anticipated future energy prices. This inherent uncertainty can be dealt with by simple studies of the effect of price variations. In such studies there is a need to be able to trade-off capital costs against operating costs; this is best dealt with using life-cycle cost analysis techniques. These techniques can then be used to determine the building design with the lowest life-cycle cost, or greatest rate of return, or whatever economic criteria is most appropriate to the person paying for the building.

With modern building technology and enough energy, almost any building design can provide comfortable conditions in most of the inhabited world. However, in a world where energy prices increasingly reflect the ultimate scarcity of fossil fuels, buildings must now be designed to be more compatible with their climate and activity. Indeed the dynamiting of large blocks of flats which were less than 20 years old[1] because they became untenable socially (crime, vandalism, etc.) could indicate the fate of some of the buildings with unacceptably high energy use that are still being constructed today.

C. Prediction Methods

Given perfect information, a building energy analysis technique which made no sweeping assumptions, and enough calculation (computing) resources, there would be no particular problem in adequately predicting the energy running cost of a building. In practice, these conditions cannot yet be fulfilled.

It is inevitable that not all the building system parameters will be well defined at the initial design stage. The designer who wishes to compare the relative energy consumption consequences of two options requires a rapid method that needs only simple input data. The designer does not have complete and detailed designs available for both options, yet these are essential for an accurate comparison. Therefore the prediction methods most useful at the initial design stage make considerable use of simplifying assumptions and default values to allow for the inherent lack of detailed design information. The question of the accuracy of the prediction, given the often sweeping assumptions made, is an area that has not been adequately considered by the design profession.

Even at the detailed design stage, it is difficult to describe the physical and operational factors of a building in sufficient depth to satisfy the input requirements of an accurate prediction method. Operational factors, such as how the artificial lights in a building will be used, are particularly hard to predict at the design stage, and yet they often have a major influence on the systems energy consumption of a building.

Thus different building energy analysis prediction techniques are subject to mutually exclusive influences. On the one hand, a simple technique will require less information and allow more parametric (altering one variable at a time) runs to be made, but with less accuracy. On the other hand, a detailed technique will model the operation of the building system more closely and should therefore give greater accuracy, but less parametric runs can be made.

Further problems with predicting building energy use are related to factors such as imperfectly known heat flow mechanisms through buildings and the use of control systems which respond much more rapidly than the currently available computer programs (with their hour-by-hour simulation of the energy behavior of the building) are able to incorporate.

There is a lack of published comprehensive comparisons of energy use as predicted at the design stage, with the energy use actually achieved in practice. This area of prediction accuracy will have to be addressed if designers are to have faith in the accuracy of predicted building energy use figures.

To some degree, the availability of computing resources puts a limitation on the accuracy of prediction methods. In most of the world, relatively easy access can be gained to large computers on which comprehensive prediction methods can be run, but at a cost. The use of telephone lines and the existence of computer bureaus allow physical access, but the cost of a number of runs can be considerable and may not be covered adequately under existing design fee scales. Certainly a case can be made, on the basis of cheaper computing, for smaller in-house computers to be used in conjunction with simpler prediction methods, potentially allowing consideration of more options. Designers have to balance modeling sophistication in a prediction method against the undoubted advantages of a more iterative design process.

D. Energy Use in Practice

There is no lack of case studies of energy use in individual buildings but systematic studies of energy use in practice are not yet widespread. This area has taken longer to produce results than the apparently more exciting areas of building energy research (such as the development of computerized energy use prediction methods). We believe that only studies of actual building energy use can, in the final analysis, establish whether certain building parameters are really associated with different energy use levels. For example, some studies[2,3] have shown that the building fabric is of less importance to the actual building energy use than is often predicted.

Using actual energy use data to examine the influence of certain building parameters one immediately finds that any group of outwardly similar buildings will have a wide range of energy use. Given a range of, say, 200 to 4000 for the AEUI (Area Energy Use Intensity[*]) for a group of "similar" buildings, does an average of 1500 actually mean much? This problem can be dealt with only by careful interpretation of the results available, and inevitably it makes definitive statements hard to support.

There is some evidence to suggest that a considerable amount of the variability in the energy use in a building is caused by the behavior of the people who rent, work in, visit, clean, maintain, and own it. Their influence is difficult to determine directly from the results of studies of groups of buildings and in many cases has been inferred from the available evidence; people's behavior is the simplest hypothesis that could explain the large degree of observed and unexplained variability.

III. PREDICTING ENERGY USE AT THE DESIGN STAGE

There is an increasing desire among building designers to predict the energy use of a building at the design stage. This would allow estimates of its energy operating cost to be made and allow design choices to be compared on the basis of their predicted energy consumption. There is no shortage of manual and computerized prediction methods. The mathematics involved are not dealt with here as they can be found in the user manuals of most large computer prediction packages or in reference books such as the ASHRAE hand-

---

[*] Although it is more appropriate to use the term Area Energy Use Intensity in the context of this chapter, the method of calculating it, the units used and the acronymn AEUI are identical to the Area Energy Use Index used elsewhere in the book.

books or CIBS guides. What is dealt with is the importance of the assumptions and simplifications that must be made in any prediction method whether it is manual or computer based.

There has been some validation both of well-documented actual buildings and of building components in the laboratory. There is still, however, a general lack of published comprehensive validation of prediction methods against actual building energy use. This suggests that these methods should be used with caution.

Energy prediction methods are very useful to the building designer. However, their inherent assumptions and limitations need to be understood if the results are to be interpreted properly and excessive scepticism or blind acceptance of their accuracy is to be avoided.

A. The Basis of Prediction Methods

Attempts to model building thermal behavior in order to size systems and components are by no means a recent development. However, since the 1973 oil crisis there has been an increased emphasis on predicting the energy use of a building at the design stage for the comparison of design options. Whether manual or computerized, these prediction techniques have to deal with three main classes of energy use determinants:

1.   The external environment (climate) of the building
2.   Its construction and systems features (physical parameters)
3.   Its management environment (user behavior)

In practice the climate, construction and systems, and management of a building are not easily separable. For convenience (as is done in the prediction methods) they are treated separately in the following sections.

The fundamental heat flow equations applying to a conditioned space are normally used to produce either an internal space temperature, or the heating/cooling space load which results from maintaining a given internal temperature. There are a number of different mathematical techniques for expressing and solving these heat flow equations. The better techniques account for the temperature (heat) flow history of the thermal paths of a space caused by thermal storage effects. To do this a family of equations is normally involved. The programs generally use either hour-by-hour real weather data or some approximation thereof as input. Then, taking internal heat (from people, lights, computers, etc.) and any latent (humidity) loads into account, the equations are solved to give the calculated internal temperature or space load.

To realistically simulate the thermal response of a building the detailed heat balance equations for each element of the structure of a building are usually calculated for hourly time intervals. The use of precalculated heat balance equations for typical constructions limits these methods to the analysis of very conventional buildings. An example of this approach is the ASHRAE weighting factor method still used in many U.S. computer programs. The use of such methods was originally adopted to minimize the computer resources and input data requirements of a building simulation but it is now recognized[4] that such methods have outlived their usefulness. The detailed heat balance approach is required in comprehensive energy prediction methods as almost every building design is different and is, therefore, a unique heat flow problem. An example of the detailed heat balance approach is the ''Matrix Method'' developed by CSIRO in Australia[5] which finds oscillatory solutions to the general heat flow differential equations. The CSIRO method uses an electrical network analogy to find the response of the thermal system to external temperatures, etc.

After calculating the space load, the more comprehensive computer programs simulate the distribution of energy in order to predict the hot water/chilled water/steam/electrical loads required to meet the calculated space load.

The program may then move into a further analysis stage where the behavior of the energy-consuming systems, in meeting the calculated load, is simulated, thus giving the predicted building energy usage (gas, oil, coal, or electricity).

Some programs also have the capability to do economic analyses. They use electricity rate structures, oil prices, equipment costs, inflation rates, discount rates, etc. to calculate the life-cycle cost of a building (the total cost over the lifetime of a building).

The assumption that the heating/cooling load calculation section can be separated from the heating, ventilating, and air-conditioning (HVAC) system operation (common to most present day prediction methods) is not entirely satisfactory in practice.[4] In fact if the HVAC plant is deliberately undersized to reduce cost and improve the part load efficiency (and hence reduce the energy use) then an indeterminate situation arises. The heat flows in the building are affected by the loads which cannot be met by the HVAC plant. At the same time the HVAC plant is responding to these heat flows. This sort of indeterminate situation can only be dealt with by iterative techniques; in fact it is usually just ignored. Similar problems arise if the building occupants can exert some control by closing blinds, opening windows, and so on. In practice, such feedback loops often arise and prediction methods are being developed to deal with them.

1. Treatment of Climate

The external climate causes heat and mass transfer through the building fabric and leads to heating/cooling loads on the building environmental services if comfortable internal conditions are to be maintained. For energy analysis purposes the external climate acting on a building system is usually represented by the temperature, solar radiation, wind speed, and humidity.

The most basic prediction methods deal only with the effect of external temperature. They assume that the building heating/cooling load is related to some measure of the coldness/hotness of the climate. A simple measure of the severity of a climate is the degree-day total for a given time period (usually a month or year). Thus if a building is situated in a climate with an annual average of 2000 heating degree days then it is conventionally assumed to require twice as much energy for heating as a building in a climate of 1000 heating degree days.

The next, more accurate measure of climate (for heating) is the bin, or temperature frequency occurrence, approach. In this method the number of hours that the external temperature falls within a certain interval (say, 5°F) is calculated for a given time period (usually a month). This method requires more calculations than the degree-day approach but the calculations are still simple enough to be done manually.

Further complex manual or simple computerized methods have been developed that take account of solar radiation heat gains and building orientation. Mitalas' Net Annual Heat Loss Factor (NAHLF) method[6] is an example of a manual method which has been developed to the point where it can give a reasonable approximation to the annual heating use of simple continuously occupied buildings.

The degree-day, bin, and similar methods are useful at the early stages of a building project and allow the designer to characterize the climate succinctly. However, these methods neglect other climatic factors, disregard the role of the building structure in heat storage, and are too simplistic in their treatment of the external temperature, to have any real chance of accurately predicting the annual energy use of a building. In areas with very cold winters the heat loss through the structure can be predicted reasonably accurately using these simple methods, which explains their widespread use in North America and Northern Europe. In areas with milder winters such methods work less well as the heat from people, lights, equipment, solar gains, and so on becomes much closer to balancing the heat loss through the building fabric.

The other end of the complexity spectrum (to manual methods) is represented by the very large computer packages that use hourly weather data. Most of these computer packages[7,8] sidestep the problem of simplifying the climate at a site. It is assumed that the hourly weather data of a given past year (or specially constructed representative year), as recorded at the nearest meteorological station recording site, is an adequate representation of the climate at the proposed site of a building. This approach has many merits but it suffers from a few still unquantified limitations, viz:

- The microclimate of the building site may differ systematically from that of the meteorological site (e.g., city heat-island site vs. rural airport site).
- Some climatic factors are usually inadequately recorded or measured, e.g., wind speed, wind direction, direct solar radiation in the path of the sun, and diffuse solar radiation.
- An hour is still too long a period for measuring some rapidly changing climate parameters such as the wind and the level of daylighting.
- The data is often incomplete, inaccurate, and only available for a very small number of locations. In the U.S. there are less than a 100 meteorological sites where solar insolation is measured,[9] too small a number for the solar radiation available in most of the U.S. to be accurately known.

The ability to quantify the climate is a prerequisite to being able to realistically model the energy use behavior of a building at a given site. The hourly characterization of climate, that has become standard practice in computerized building energy prediction techniques, represents a good balance in most respects between data requirements, cost of computing, and accuracy. It should be remembered, however, that accurate modeling of some aspects, such as control system responses (in complex HVAC systems), and natural ventilation rates, are best done at shorter than hourly time intervals.

2. Modeling the Physical Parameters

For modeling purposes the building physical parameters can be divided into two groups, the building construction and the building systems (environmental and utility services).

B. Construction

The construction of a building is a prime determinant of its energy use because of heat and mass transfer through the building fabric plus storage effects due to its mass. In all but the simplest buildings the accurate modeling of the construction will be a very complex problem due to the large number of different materials usually involved and the four-dimensional nature of the problem, three spatial dimensions for heat and mass transfer, and time for storage effects. Since rigorous modeling of this situation is impossible, two important simplifying assumptions must be made.

The first assumption is that heat conduction through the fabric is one dimensional. In the manual prediction methods this is further simplified to a one-dimensional steady-state heat transfer plus, sometimes, a simple heat storage factor.

The second assumption is that the thermal and air infiltration aspects of the building are adequately known and modeled and can be consistently described by the users of the prediction method. This aspect of interpretation can be very important.

Different comprehensive computer prediction packages usually give very dissimilar estimates of energy use for a given building design. Figure 1 shows the results obtained by different computer programs for a given simple commercial test building. This building simulation did not include the effects of ventilation, infiltration, and HVAC systems. The simulations were performed by highly skilled people familiar with the packages used. ''Since substantial effort had been made to ensure that input assumptions about building character-

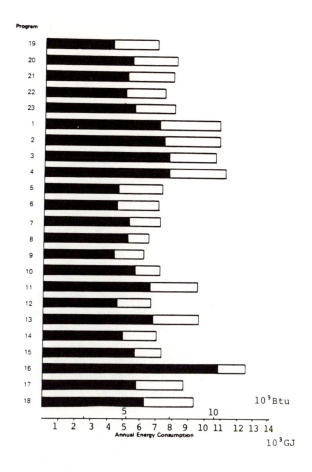

FIGURE 1.   Predicted heating and cooling energy consumption for different prediction programs. No ventilation or infiltration and no HVAC plant simulation were included. *Note:* Energy consumption based on calculated annual heating and cooling loads assuming a heating plant efficiency of 100% and a chiller plant coefficient-of-performance (COP) of 3.0.

istics were essentially the same, differences in program results can be attributed primarily to differences in the analytical methodologies and assumptions built into the different programs."[10]

Even when different analysts use the same simple prediction technique and the same building data, it has been found that their differing interpretations of the building construction can lead to significantly different predicted values.[11]

C. Services

The services of a building comprise all the energy-consuming equipment used to heat, cool, ventilate, light, and transport the occupants. The main assumption made in prediction methods about the environmental services and the building internal heat gains (both sensible and latent) is that their characteristics are known and can be realistically modeled. This assumption is questionable even when all the environmental services are functioning correctly; but when, as often happens, maintenance is neglected, then the actual services behavior may be very different from that assumed. The internal loads due to such factors as the heat and moisture produced by the building occupants is usually not well known at the design stage either. The addition of extra equipment to a building during its lifetime can lead to high energy consumption or substandard comfort conditions that cannot always be allowed for in the original building design. For instance, a well-designed building may have a high energy use if large computer systems are added by tenants, leading to continuous high air-conditioning loads.

While the mathematical modeling of heat transfer through structures such as large heavy masonry walls is usually quite adequate, modeling the many other heat and mass transfer processes involved in real buildings presents some difficulty. The relationship between wind and air infiltration is ignored in most available models, as is the effect of the wind on the external heat transfer coefficient of components such as windows. Other examples of significant thermal aspects that are usually inadequately dealt with are thermal stratification and stack effects, the availability of daylighting throughout the building, heat transfer between zones with wide thermostat dead bands, and moisture storage in the structure.[4]

1. Modeling User Behavior

The two previous sections have dealt with the general problem of modeling a building situated on a given site, subjected to real weather, constructed with unknown skill, of materials that may differ in their properties from those found in the laboratory, and which respond in unknown ways. To this situation must be added a further complicating set of factors, the behavior of the people who will manage, maintain, occupy, and visit the building during its lifetime. Although the initial building design is undoubtedly a very important factor in the actual energy consumption of a building, the effect of energy management and occupant-related factors must not be underrated.

The study of the Twin Rivers housing estate in New Jersey is one of the most comprehensive occupant-related energy studies so far published.[12] The Twin Rivers estate consisted of several hundred essentially similar single-family houses occupied by young, middle class, white professionals. It was found that the energy consumption of similar houses could vary by a factor of two in an essentially random manner. The major role of building user behavior is highlighted by this study.

The random nature of user-related factors means that even with all the physical parameters specified it would be foolish in the extreme to predict the energy consumption of a building. One can talk only of expected averages. It is optimistic even to expect that the predicted relative savings of one design feature vs. another will hold. The actual building operation can easily reverse the advantages predicted for one option compared to another. For instance, it has been found that New Zealand hospitals with central coal fired boilers supplying hot water to radiators are much higher energy consumers than electrically heated hospitals.[13] This is due to the users (staff and patients) opening the windows to reduce heating rather than turning down the radiator valves. Opening windows is more obvious and leads to more rapid results but obviously wastes energy; the heat from the radiators goes straight out of the windows.

D. Using Prediction Methods in Design

Methods of varying complexity can be used to predict whether a given building will meet some previously defined design criteria. The designer may well have several choices of building design, each of which could adequately meet these criteria. For each design choice the expected annual energy and maintenance cost could then be compared with its initial capital cost, in a life-cycle analysis. Life-cycle analysis methods allow costs to be compared with benefits even when they occur at different stages in the lifetime of a building.

A life-cycle cost analysis may show that with double glazing, reduced energy use benefits during the lifetime of a building would lead to a higher profit for a property developer borrowing money at commercial rates. Sensitivity analysis may show that double glazing is still cost effective with higher interest rates and lower fossil fuel prices. In this case the developer would know that double glazing was a robust option worth considering seriously.

Sensitivity studies are seldom used when deciding between options. In a world where the future is unknown this is too simplistic ("Those who foretell the future lie, even if they tell the truth" — Middle-Eastern proverb). Only by examining a number of future scenarios can the designer be confident that the advantage one option has over another will hold even if future parameters differ slightly.

The uncertainty of predicted energy use values limits their usefulness in the making of design decisions. A study by the Solar Energy Research Institute addresses these limitations: "It is clear that caution must be exercised when using the absolute energy consumption calculated by a single BEAS (Building Energy Analysis Simulation). The practice of using results from a single BEAS to generate design guidelines or simplified tools or input for economic analysis or to develop policy, is questionable."[14]

Unfortunately in a world of finite resources of money and manpower, designers cannot perform detailed life-cycle cost/benefit analyses on all the major energy-related design options, let alone tests of sensitivity. The choice is between a large number of simple simulations or just a few detailed simulations.

A way out of this dilemma of complexity vs. simplicity would be to have an optimizing program that picked the best overall solution. Unfortunately, optimizing programs would require massive amounts of accurate input data and would be very costly in their use of computer resources.

E. Predictions Based on Empirical Data

A rather different approach to predicting energy use at the design stage is to use the energy consumption history of a group of buildings located in a similar area to the proposed new building. This approach seeks to:

- Make the intuition (based on experience) that most designers use more explicit
- Avoid many of the modeling problems discussed in the previous sections
- Utilize the real data of actual buildings (operated by normal imperfect people)

The Building Services Research and Information Association[15] (BSRIA) used this type of approach during a study of the systems energy consumption of 626 buildings in the U.K. Targets were derived for offices, factories, warehouses, schools, shops, and hotels based on the lower 25 percentile energy use intensity of the particular building type. For example, the BSRIA report suggests a target building systems AEUI of 600 MJ m^{-2} year^{-1} (53,000 Btu ft^{-2} year^{-1}) for well-insulated, maintained, and operated office buildings with good heating controls. This suggested energy target may be modified for different building parameters. For instance, the target is multiplied by 1.4 for air-conditioned or mechanically ventilated office buildings.

Targets arrived at, using this empirical data-based approach, can only be a first approximation. Because they are based on what other similar buildings have achieved, such targets sidestep the problems related to prediction methods and give the building designer and also the building operator a rough estimate of the expected energy use. This estimate is useful in the same way that in the U.S. the Environmental Protection Agency (EPA) fuel economy figures for automobiles are useful, viz., for comparing different vehicles before a final choice is made on which model to buy. Such targets give a yardstick of the expected performance against which the actual performance can be measured. If an automobile with an EPA (or other) highway fuel economy figure of 30 mi/gal (8.5 ℓ/100 km) is only achieving 15 mi/

gal (19 ℓ/100 km) in highway use, then the driver can be fairly sure that something is very wrong. Similarly, if a building has an AEUI of 2000 when similar well-maintained buildings have AEUIs of 1000 then one can be reasonably sure that the building is energy inefficient.

A commentary by Ternoey[16] entitled "Don't Trust Your Instincts: Big Building Design Defies Intuition" addresses this need for historical energy use data, stating that "until we have easier design methods or rules of thumb, we should start each commercial project by dissecting a similar building." Unfortunately, there is insufficient published data on systematic studies of energy use in groups of buildings to enable this "dissection" to be readily undertaken.

In areas where new technologies are still being developed, techniques based on empirical data cannot be developed entirely from existing buildings. The techniques may, in addition, include the results of laboratory tests of individual components or modules. Such tests can be used to validate computer-based predictions of performance. One area where this approach has been used recently is that of solar-conscious design. Here the historical performance base was weak and hence design methods such as F-Chart and Solar Load Ratio gave designers a starting point. Despite the limitations outlined earlier, such techniques are very useful. Given the rapidly moving state-of-the-art in areas such as solar design, these initial rules of thumb will have a short useful life before they are replaced by better prediction techniques based on empirical data.

Prediction methods based on empirical data are most useful for considering broad system parameters, since, at the detailed level, cause and effect become difficult to separate in a population which is usually made up of one-off buildings. The provision of reasonable "ball-park" figures at the initial design stage can enable the basic decisions on broad system parameters to be made with some idea of their likely future energy use consequences. This allows scarce resources (money and manpower) to be concentrated in the areas of greatest energy conservation potential.

IV. ENERGY USE IN PRACTICE

This section examines the relationships found in actual buildings between systems energy consumption and a number of broad system parameters. The examination is deliberately limited to measured building energy consumption in normal commercial and institutional buildings, as the utility of predicted energy consumption figures rests on a number of questionable assumptions (discussed in the previous sections). In the final analysis, theories concerning the importance of certain factors in determining the energy use of a building must be tested by examining the energy use characteristics of normal buildings. There are a great number of published case studies of energy use in single buildings but these are of limited use in determining the systematic importance of certain building parameters. This chapter uses only the results of the published systematic studies that we could find.

These studies tend to agree on a few general conclusions, the most important of which is that there is a large diversity in energy use intensity, both for large groups of buildings and for smaller samples of buildings with similar activities.[2,3,12,17,18] Good design, coupled with good building maintenance and operational practices, can result in very low-energy-use buildings indeed, often many times smaller than average. There is, however, almost no limit to how much energy can be wasted (perhaps the accountant's willingness or ability to pay the energy bills is the ultimate limit?) through inappropriate design, maintenance, or operation of a building. This tends to explain the wide spread of energy use intensities observed in practice, ranging from a 2:1 variation at the Twin Rivers housing estate[12] (for very similar houses) to the 10:1 variation found in buildings housing similar commercial activities.[2,3,17,18]

Unfortunately, attempts to link cause and effect, in determining the reasons for the large observed variability, come up against the fact that most building parameters are highly interrelated. In contrast to computer predictions, in the real world it is very hard to maintain the required experimental control of all the other factors in order to assess the effect of varying a single parameter. For example, it is likely that a building of a certain style with a very high glazing ratio will have a large and complex HVAC system, for reasons also related to that particular style. If this building had a high-energy-use intensity, could it be said to be caused by the high glazing ratio, by the large complex HVAC system, by the building style, or by the nature of the exogenous factors which led to that particular style being adopted?

Because of the complex interrelationships between building parameters it is very difficult, for example, to justify the statement that central HVAC plant causes high-energy-use intensity, but some studies have shown that central HVAC plant is associated with buildings having a higher than average AEUI.[2,3] Several studies have shown that the larger type of office building, with central HVAC systems, has a high average AEUI.[2,3] However, the AEUI was not found to be significantly different from that observed for buildings that did not have central HVAC systems. In our experience, differences in observed AEUI for single building parameters are usually not statistically significant (in studies of actual energy use in nonresidential buildings).

The use of average energy use figures for groups of buildings hides the large spread about the average value that usually exists. Where figures for this variation are published, it is found that the standard deviation is of a similar size to the mean. The significance of any differences must be statistically tested before anything can be concluded from such figures.

Actual results are often unpalatable in that factors usually predicted to be very important in building energy use are found to have little systematic effect. The discovery that some "low energy" designs[19] have not achieved their design energy performance should inject a needed cautionary note into the optimistic claims of energy performance which are all too common.

A. Time Varying Influences

The energy use characteristics of buildings have changed over the years, as a result of advancing technology, evolving building codes, changing architectural styles, and a myriad of other reasons. As well as the influence of the year of construction and the physical aging of equipment, there is an aging effect which refers to the changes in the energy use characteristics of a building resulting from improved understanding on the part of the operator (a learning curve situation). There may also be changes in activity and alterations in user expectations during the lifetime of a building. This changing of the energy use of a given building with time is important for designers who are trying to evaluate the success of their designs. Only when the designer's assumed pattern of operation is shown to be valid can a meaningful comparison be made between actual energy use and energy use predictions made at the design stage. We shall now consider time varying influences under the three headings: year of construction, effect of climatic variation, and building aging.

1. Year of Construction

The past century has seen major changes in architectural styles and correspondingly important developments in the energy-consuming equipment used to enhance the building occupants' comfort (exclusion of noise and dust, temperature and humidity control, lighting, and transportation). In most industrialized nations commercial buildings have changed from heated, naturally ventilated, massive masonry construction with low levels of artificial lighting to "modern style" steel or reinforced concrete structures with lightweight cladding, often with full air conditioning and high installed artificial lighting loads. In many cases these changes have been strongly influenced by major developments in building codes and standards.

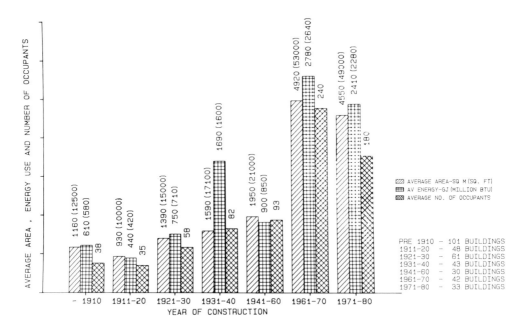

FIGURE 2.   Area, energy use, and number of occupants grouped by year of construction (Wellington CBD).

The question of interest is to what extent have these changes affected energy use characteristics? Were buildings constructed in the era of very cheap energy (1950 to 1970) energy wasteful? In our view, based on the available evidence,[2,3] there seems little reason to believe this to be the case. Newer buildings tend to be larger and use more energy but the ratio between size and energy use is fairly constant. Figure 2 for the Wellington CBD illustrates this trend clearly, the period 1930 to 1950 being unusual by virtue of the early 1930s Depression and the 1939 to 1945 World War. The Hittman study[3] of a group of buildings in the Baltimore CBD came to similar conclusions: "Thus, it seems that the main conclusion to be drawn from the observed lack of correlation, is that variation in building age does not imply any sufficiently strong systematic variation in any of the meaningful thermodynamic variables to influence the EUI in an observable manner. Put more simply, age seems to be a poor surrogate for any single variable, or group of variables, which substantially influence EUI."

Given that building age seems to be very weakly linked (if at all) with the resultant AEUI of a building, it may be argued that more recently constructed buildings provide more uniformly comfortable conditions for their occupants, at no extra cost. There are serious doubts, however, that uniformly "comfortable" conditions may be as desirable as has been assumed (Chapter 3).

2. Effect of Climatic Variation

Most design calculations make the assumption that the energy use of a building will be closely related to the severity of the climate. This climate dependence should be observable within the changing seasons of a given year. The varying climate at a given location from year to year should also lead to differences in energy use between years. Finally, similar buildings in different locations should show energy use differences that could be related to climatic variations. Figure 3 shows the relationship between average heating consumption (natural gas) and average monthly working day temperature for 157 Wellington CBD buildings in 1978. This performance line (see Chapter 7 for a more detailed discussion of performance lines) indicates an essentially linear relationship between monthly heating fuel

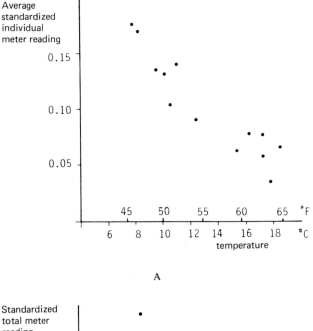

A

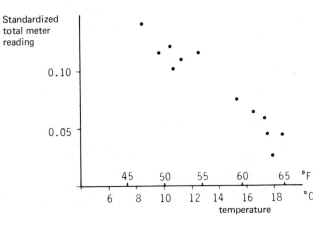

B

FIGURE 3. Temperature elasticity of gas consumption for 157 buildings. (The meter reading for each period (1 month for gas) has been standardized by dividing by the total annual consumption. In (A) the standardization is performed for each meter then the average of all these standardized readings is calculated; in (B) the sum for all the gas meters is calculated before the standardization.) (A) Average individual standardized monthly gas meter reading for 1978 vs. average monthly workday outdoor temperature (illustrates behavior of smaller commercial buildings). (B) Standardized total monthly gas consumption readings for CBD commercial buildings in 1978 vs. average workday monthly outdoor temperature (illustrates behavior of larger commercial buildings).

use and outside temperature. Both small and large users of gas (Figures 3A and 3B, respectively) in the Wellington CBD appear to respond to the average monthly temperature in a similar fashion.

In the Wellington CBD, electricity use was found to be slightly temperature dependent for small consumers, while for larger consumers little dependence could be found. The small consumers were more likely to use electricity for heating so this result was expected.

It can therefore be concluded that in both large and small Wellington CBD buildings, heating fuel use varies in response to the average outside temperature. One would expect that annual variations in coldness would be reflected to some considerable degree in heating fuel use in commercial buildings. This has so far proved difficult to observe in practice, however, due to the influence of other trends such as energy supply constraints.[20] Table 1 shows the relationship between heating degree-day totals and the average AEUI for groups of buildings in a number of different commercial sectors. It can be seen that there is a trend towards higher average AEUIs in the more severe climates. The figure for New York is higher than would be expected but many buildings there are very large and contain features which are not common in buildings in other cities, for example, steam heating and absorption cooling. Figure 4 shows the very strong relationship between AEUI and annual heating degree days for groups of New Zealand and North American schools.

The relationship between the cooling energy use of a building and the hotness of the climate (e.g., cooling degree days) is more difficult to establish than that of heating energy use. This cooling is usually provided by electrically powered chillers but the electricity use is often not metered separately from other electricity end-uses, such as the power to drive the ventilation fans. In some climates it is possible to cool the building mass at night by using colder nighttime air, especially in desert regions where comparatively cold nights are common, even in the middle of summer. It may also be possible to employ evaporative cooling which is relatively cheap if the water is available. To date, little systematic data has been published on nonresidential building cooling energy use. This is an area ripe for examination because of the considerable demand for cooling in many parts of the world.

3. Building Aging

To some degree a building resembles a living organism in that energy must be continuously expended to keep it operating. There is a constant battle between order and decay. It takes considerable effort to maintain order in a building over its physical lifetime, an effort which requires interrelated elements of intelligence, capital, and energy.

If maintenance (comprising intelligence and capital) is allowed to slip there will usually be subsequent energy use penalties. Anyone familiar with buildings will almost certainly have experience of the common results of inadequate maintenance, for example, outside air dampers that do not move, blocked air filters, time clocks not changed for daylight saving, chillers and boilers running simultaneously. The consequences of years of inadequate maintenance can be enormous; in the U.K. the Property Services Agency achieved an overall 35% energy savings in Government buildings over a 6-year period by concentrating mainly on remedial maintenance and the upgrading of controls.[23]

When examining the effects of building aging in recently constructed buildings one often finds a situation where the learning process dominates the energy use. Figure 5 shows the use of gas for heating during the first 5 years of full occupation of an office building in Wellington. The 14% reduction between 1978 and 1979 was achieved by a keen building manager (without specialist knowledge) who tuned the building heating system with existing controls and sensors.

By 1982 a computer control system had reduced the gas use by 46% compared with 1979. The computer system also operated 14 other buildings and provided an energy and security management system.[24] The change to external computer control involved further fine tuning and additions to the original building controls system. The internal building comfort conditions have not been compromised in making these reductions.

This example vividly illustrates the learning curve process for a new building and raises the important question of which year best represents the energy performance of the building? After all, the curve for 1981 or 1982 could probably have been achieved by a well-educated and motivated building operator with the original building controls. The answer to this

Table 1
MEAN AREA ENERGY USE INTENSITY FOR COMMERCIAL BUILDINGS IN FOUR CITIES

| CBD | Mean AEUI | | Heating degree days | | Av area | | Number of buildings in sample |
	MJ m⁻² year⁻¹	Btu ft⁻² year⁻¹	°C days for 18°C base	°F days for 65°F base	m²	ft²	
Auckland[17]	570	50,000	1,179	2,122	650	7,000	770
Wellington[2]	620	55,000	2,074	3,733	1,410	15,200	940
Baltimore[3]	1,134	100,000	2,585	4,653	5,280	56,800	243
New York[18]	1,340	118,000	2,706	4,871	37,300	402,000	44

Note: These AEUI figures are based on building boundary energy use data. The AEUI is the total energy use divided by the total sample gross floor area.

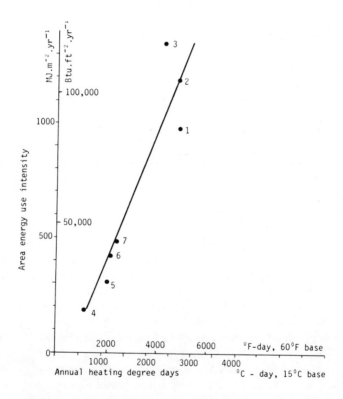

FIGURE 4.   AEUIs vs. degree days for schools (1) New York City oil-heated schools[21] (2) New York City coal-heated schools;[21] (3) Fairfax County, U.S.;[22] (4) Auckland, N.Z.; (5) Wellington, N.Z.; (6) Christchurch, N.Z.; (7) Dunedin, N.Z. The coefficient of determination (r²) = 0.88. (Results for New Zealand based on samples of approximately 20 schools for each locality.)

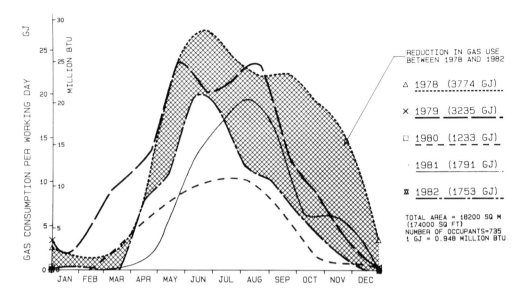

FIGURE 5.   Gas reduction due to improved management for a new building — located in the Wellington CBD.

question is very important if the energy performance of a building is to be compared with a target value.

A new building will usually have a settling in period. Only when this period has elapsed will it prove useful to compare the energy performance with that of similar buildings or with the predicted values. In an older building, where tired equipment is wasting energy, the internal conditions may be quite tolerable. One cannot rely on the common operating philosophy that if the plant is still running and the building is reasonably comfortable then all is well. Such a building may be providing those conditions at a much larger energy cost than is necessary.

B. Effect of Building Form and Fabric

In this era of increased energy awareness most building designers want to make a contribution to energy efficiency, and the building form and fabric is an area where their concern often manifests itself. This concern may be at the trivial level where symbols associated with energy-conscious design are used to create an aura of energy efficiency: identical sun shades on all the windows of a building when only those facing the equator are really effective, for example. More often the concern is manifested in features thought to be related to energy efficiency, which are, however, based on the dissimilar and very personal experiences of domestic construction. An example of an intuitively "obvious" energy-efficient feature is double glazing which may not necessarily lead to much energy use reduction, especially in buildings with high internal heat gains. Sometimes, the form and fabric are skillfully combined with the function, climate, and services of the building to deliberately and successfully create a low-energy-consuming building. Such integration is, however, not common.

A major source of information on the energy consumption effects of building form and fabric are published design guides, usually based on computer simulations of "prototypical" buildings. The anticipated benefits are often not fully realized in normal buildings because of the limitations of this approach (discussed in more detail in the previous section of this chapter).

Concentration on a number of building design features can result in amazingly low-energy-

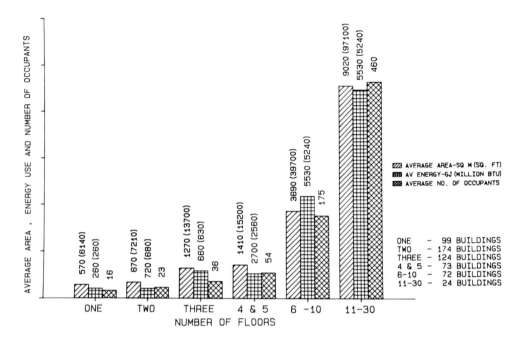

FIGURE 6.   Area, energy use, and number of occupants grouped by number of floors (Wellington CBD).

use buildings. An extreme example is St. George's County Secondary School, Wallasey, Cheshire, built in the U.K. in 1961, which requires no auxiliary heating[25] (though at the cost of compromised performance in other areas). However, such extremely low-energy-use buildings do not always have optimal flexibility of operation, and the capital costs in both money and energy may be high. Given finite design resources, too great an emphasis on form and fabric will divert scarce resources from areas such as control systems and probably lead to a building which uses more energy than would result from a better integrated design.

1. Form

The heat loss or gain of a building is to some degree related to its exposed surface area. It is often suggested that a compact building form will serve to minimize these heat losses or gains, that is, reduce the effect the external climate has on internal thermal conditions. This point of view underrates the positive contribution to internal conditions that the external climate can make via an appropriate climatic filter (the building fabric). A building designed to take advantage of passive solar heating and cooling and also provide good natural daylighting would not have the compact hemispherical shape (such as an igloo) theoretically best suited to minimizing heat loss and gain. This is because the building must face the sun for passive solar techniques to work and it must be narrow for daylighting to be most effective.

For the building researcher, the most readily available factor relating to the compactness of a building is its height, as measured by the number of floors. Taller buildings tend to depart more from the "ideal" compact shape than low rise buildings, because of constraints on base area due to zoning regulations and available site sizes in inner city locations.

Using the number of floors as a measure of building compactness, it is possible to examine its relationship to energy use in large groups of buildings. Figure 6 shows the relationship between energy use, building area and number of occupants, and the number of floors for approximately 560 buildings in the Wellington CBD. The AEUI does not significantly differ

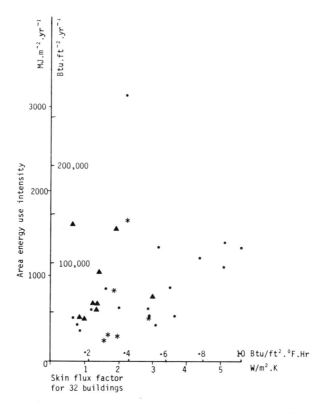

FIGURE 7.   Energy use intensity variation with skin flux factor (for 32 general offices in Baltimore). *Note:* skin flux factor = (average area weighted U value of exterior glass, walls, and roof) × (exposed skin area/ gross floor area). *, Both packaged and central systems; ▲, central systems; ●, packaged system; r^2 = 0.643 for packaged system (n = 15); r^2 = 0.895 for central system (n = 6); (r^2 = coefficient of determination).

(at the 95% confidence level) with the number of floors in these buildings. This is not to say that a correlation between building compactness (height) and energy use does not exist, but no significant correlation could be observed in this study. The results of the Hittman study[3] of several building activity groups also found no significant relationship between average AEUI and building height (a measure of compactness). It is clear that the main effect of taller buildings is that they are, on average, larger and hence use more energy.

A sample of 140 larger buildings (mainly offices) in the Wellington CBD was studied in considerable detail to examine the energy use significance of form related factors. The factors examined were the exposed external surface area, the ratio of glass area to this exposed external surface area, and the compactness ratio (as measured by the ratio of the exposed external surface area to the surface area of a hemisphere of the same internal volume). In multiple linear regression studies, these factors were found to make no significant contribution to "explaining" either the energy use or energy use intensity (AEUI) of these buildings.

There have been some studies[17,18,26] that have found a small significant correlation between building form and energy use. However these studies were for comparatively small numbers of buildings and the authors have pointed out that the results should not be extrapolated to larger groups of commercial buildings.

There are many examples of buildings[25,27] where the architectural form has played a vital role in providing energy-conserving features such as enhanced daylighting to reduce electrical lighting energy use, earth berming to reduce the effect of the prevailing wind, minimization of external surface area to reduce heat loss, placement of glazing to maximize winter solar gain and reduce heat loss, etc. Based on the results of individual buildings, there seems little doubt that the building form can be a vital element of low energy use in buildings, but it must be an integrated and deliberate part of a good overall design to make its full contribution. Variations in form among existing commercial buildings have not been found to significantly affect energy use.

2. Fabric

For buildings with low internal heat gains (residential buildings, for example) there is little doubt that both the thermal resistance and capacitance (heat storage) characteristics of the fabric are almost always very important parameters in heating and cooling energy use. For commercial buildings with high internal heat gains, the heat loss and heat gain through the fabric assumes less importance and may have a smaller influence on the heating and cooling energy use than such parameters as solar gain, ventilation and infiltration, lights, occupant heat loads, and other loads (for example computers) as well as the HVAC plant characteristics.

Figure 7 shows that for a group of 32 commercial buildings in the Baltimore CBD[3] there is little observable direct relationship between energy use per unit area and a calculated overall skin flux factor (an area weighted U-value). Analysis of 450 Wellington CBD commercial buildings[2] also failed to find any appreciable association between energy use and whether the building had light or heavy construction walls and roofs.

Certainly there are individual low energy buildings where the building fabric plays a vital role, but this is more due to good integrated design than the influence of the building fabric alone. Simulations of individual buildings or "typical" buildings usually establish the theoretical importance of building mass and thermal resistance, but this importance has not been verified in practice for individual cases in groups of buildings. To further complicate matters there is some evidence that the thermal properties of building materials are different when measured in actual buildings rather than in the laboratory. Some of this variation is probably due to varying material properties, but a large amount of uncertainty is added as a result of differing levels of skill in construction, leading sometimes to major thermal short circuits. The airtightness of the fabric is also of great importance. In a windy location, very high rates of natural infiltration may be observed, high enough in some cases to lead to discernable internal air movement and subsequently high heat losses.

C. Influence of Energy-Consuming Services

Much has been written about the theoretical behavior of the energy-consuming services in buildings. Given the enormous variability in and complexity of the lighting, transport, and HVAC systems of most nonresidential buildings only the effects of broad system parameters can be determined in most real world studies. Comparisons of the results of using different HVAC system types are difficult to obtain in a form that is likely to apply to buildings in general. Our objective here is to illustrate the areas where the choice of certain system types is associated (or not) with significant AEUI differences. For instance, certain combinations which many designers believe to be inefficient can be shown to have somewhat higher AEUIs in practice too. A good example is large computer installations in buildings with central HVAC plant systems; see Figure 8.

Attempts to show that, say, higher installed lighting loads lead to greater energy use are often frustrated because of the overwhelming effect the hours of use, switching layout, behavior of cleaners, and building management have on actual lighting loads in normal buildings. In fact it is possible to find buildings with a great variety of energy-consuming

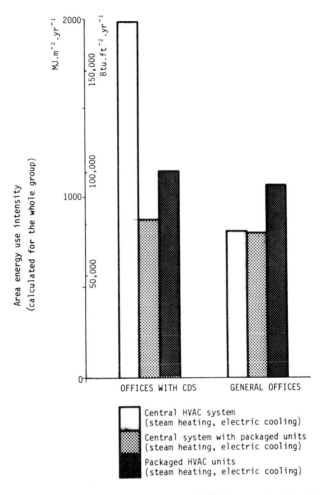

FIGURE 8.   Area energy use intensity by HVAC system for offices in Baltimore with and without CDS Facilities.[3] (CDS, Computers, data processing, and support facilities.)

system types, both among the lowest and the highest energy use categories. System features that are expected to cause high energy consumption often have other subtle implications that may lead to low energy use if managed appropriately.

In practice there is a considerable incentive to keep building environmental services as simple as possible. Complexity in building environmental services often leads to situations where a supposedly superior option does not demonstrate its expected advantages due to human factors (mentioned in more detail in the following section) or inadequate controls. Equally, an overemphasis on simplicity of operation will often lead to energy waste through inflexibility; the designer should try to strike a balance between crude simplicity and incomprehensible (to operate) complexity.

1. Ventilation Method

The method used to ventilate a building usually narrows down the subsequent choices for the rest of the HVAC system. The choice of ventilation method is usually strongly affected by various exogenous factors. These factors include the size of the building, its external climate (especially wind, outside air temperature, and humidity) and the external noise and air pollution levels.

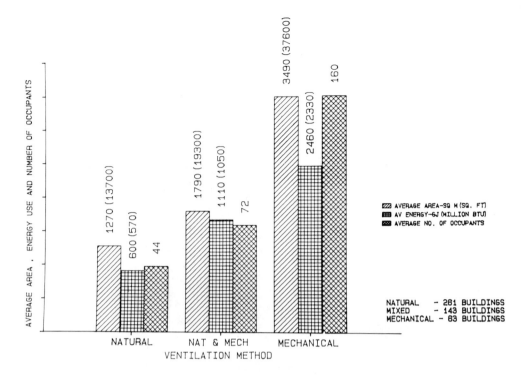

FIGURE 9.   Area, energy use, and number of occupants grouped by degree of mechanical ventilation (Wellington CBD).

In situations where it is a viable option, natural ventilation using windows is assumed (see Chapter 3) to lead to lower energy use. This has led to a renewed interest in natural ventilation in milder climates. Figure 9 shows the average relationship between building size (as measured by floor area and number of occupants) and building energy use, for 460 commercial buildings in the Wellington CBD with different levels of mechanical ventilation. It can be seen that mechanically ventilated buildings tend to be bigger, to consume more energy, and to house more people. The actual average intensity of energy use per unit area or building occupant does not increase systematically with increasing centralization of the ventilation system of a building. The results of regression studies[2] support the conclusion illustrated by Figure 9 that the ventilation method employed in a building is not highly associated with the AEUI or PEUI.

There are a number of possible reasons for this result. Naturally ventilated buildings are more likely to be built with heavy emphasis on low capital cost. They are also more likely to contain energy-intensive activities such as restaurants or saunas. Wellington's windy climate is perhaps rather hard on naturally ventilated buildings too. An average wind speed of 7.5 m sec^{-1} (16.7 mi/hr) is certainly very high and could easily lead to excessive ventilation and hence heat loss for a large part of the year. The results of the BSRIA study[15] in the U.K. illustrate that although mechanically ventilated buildings use more electricity, they also use less fossil fuel than naturally ventilated buildings. This seems reasonable in view of the electrical requirements for fans in mechanically ventilated buildings; the lower fossil fuel use is most likely due to the recirculation of air.

In well-insulated domestic buildings the ventilation heat loss usually becomes the major heat load. There is a trend in very low energy domestic construction in cold climates to reduce infiltration and ventilation to the minimum that is technically feasible. The ventilation air is then provided using air-to-air heat exchangers which use very little energy. For

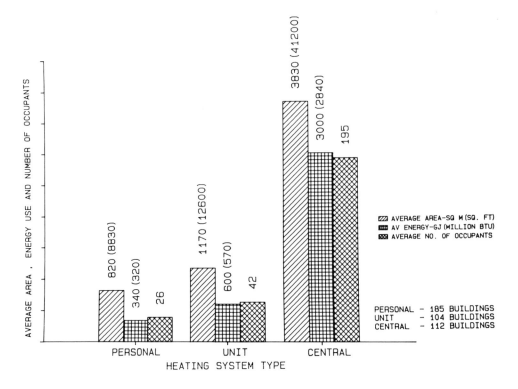

FIGURE 10.   Area, energy use, and number of occupants grouped by heating system centralization (Wellington CBD).

commercial buildings, a similar situation can apply. The potential to reclaim energy from the exhaust air and to cool the structure with cold outside air at night, can in many cases justify the expense, complication, and fan energy use of mechanical ventilation.

2. Centralization of Services

The centralization of building services presents certain inherent advantages and disadvantages to the building designer. It is one of the critical early design choices that must be made and has far reaching effects.

a. Advantages

There are often considerable pressures for centralized building services, especially for heating and cooling plants. The arguments are based on apparent economies of scale. It appears cheaper and generally easier to provide one or two large items of central plant than to have many smaller units distributed throughout the building or complex. The average efficiency of the central plant can be improved by having several units (often of different capacity) and then operating the individual units closer to their most efficient operational point. There is also the potential to have spare units available to relieve the pressure on maintenance and to improve the overall system availability (percentage of time the system is available for operation). Finally, advantage may be taken of the cheaper fuels which large centralized boiler installations can more readily utilize.

b. Disadvantages

Centralization of services has disadvantages too. These chiefly relate to complexity of control, inflexibility of operation, and the difficulties of distributing heated/cooled fluids

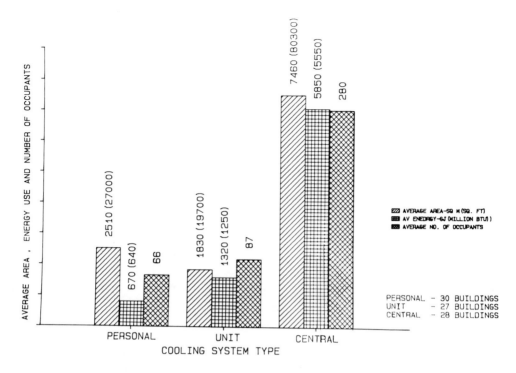

FIGURE 11. Area, energy use, and number of occupants grouped by cooling system centralization (Wellington CBD).

(usually air or water) efficiently. With centralized systems, flexibility of control is difficult to achieve without considerable additional complexity and cost, and even then it is rarely possible to operate a single office outside normal working hours without running the large HVAC units of the whole building inefficiently. Centralized control of lighting can lead especially to problems with use outside normal hours; it is not uncommon to find all the building lights on while cleaners are working, causing significant extra lighting energy use.

Evidence from empirical studies of building energy use, on the effect of centralization of services, is not very conclusive. Figure 8 is interesting because it shows lower energy intensity for the general offices with central HVAC systems (possibly reflecting their higher inherent efficiency) compared with packaged HVAC systems. The presence of computers in buildings with central HVAC systems is associated with higher energy use intensity, probably due to the 24 hr/day cooling and tight humidity control required by some computers leading to central HVAC systems running almost continuously at low overall efficiency. Figures 10 and 11 for buildings in the Wellington CBD show that the average AEUI or PEUI is not systematically linked to the centralization of heating or cooling systems. It would seem that the real world "noise" drowns out any systematic trends.

3. Lighting

Electricity consumed by artificial lighting systems is often a major component of the total energy use within a building. The heat output of lights can be a significant contributor to the internal heat gain of a building, leading to either a useful source of warmth or an unwelcome source of overheating, depending on conditions at the time. It is now widely recognized that some of the extremely high lighting levels provided in an era of cheap electricity led to buildings with high cooling loads.

In nations where electricity is largely generated from fossil fuels with a subsequent ratio

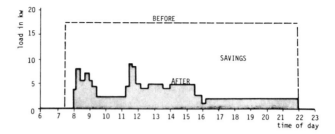

FIGURE 12. Lighting load reduction by altering switching for the Lester Pearson Office Building in Canada. (From Hunt, D. R. G., *Build. Environ.*, 14, 21, 1981. With permission.)

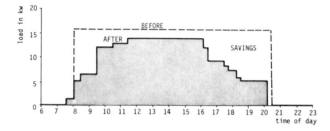

FIGURE 13. Lighting load reduction by altering switching for the National Library Archives Building in Canada. (From Hunt, D. R. G., *Build. Environ.*, 14, 21, 1981. With permission.)

of primary energy to delivered energy of around 3:1, lighting will often dominate the primary energy use of a commercial building. The price of electricity often reflects its high primary energy component. Electricity is generally more expensive than fossil fuels, making artificial lighting a larger fraction of the annual energy cost of the building than its proportion of the delivered energy of the building would initially suggest.

Lights only use energy when they are switched on, an obvious statement perhaps but it is often overlooked. However, much of the available literature has concentrated on the installed loads of lighting and neglected the "hours of use" aspect. Lighting energy use is the product of the actual lighting load and actual hours of use; both are equally important. Figures 12 and 13 show the energy use savings due to reductions in the number of hours individual lights were used per day. These savings were achieved by a change in the light switching arrangement; note the installed lighting load is unchanged! Figure 14 indicates a weak relationship between installed lighting levels and annual electricity use for 13 Baltimore CBD buildings; again it would seem that the hours of use are as important as the installed lighting load.

Good lighting design is a specialist field, but it would seem that insufficient thought is given to the needs of all the people who will use the lights and how the lights will be controlled. The current interest in task-ambient lighting is a case where care must be taken to provide adequate levels of general illumination in public areas and for cleaners if the expected low-energy-use benefits are to be realized in practice.

The use of uniform lighting layouts, with no consideration of variations in illumination required, and with central switching, is becoming less justifiable. Given the frequent high cost of electricity, it is increasingly desirable to design lighting layouts which supplement the available daylight and provide good control over a small number of luminaires. There is a place for simple direct controls, such as pull-cords, on individual luminaires for direct and unambiguous user control. Photocells to turn off or dim lights when adequate daylighting

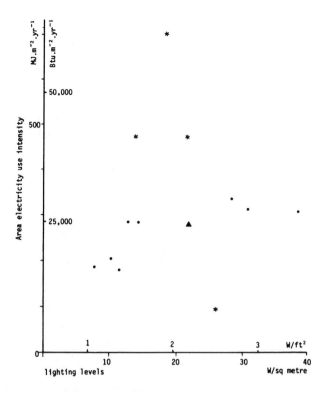

FIGURE 14.   Annual electricity usage by lighting levels (for general offices with steam heat and electric cooling). *, excluded from calculation of r² value; ▲, central HVAC systems; ●, packaged units or combined systems. ▲ and ● the coefficients of determination (r²) = 0.64 for these nine buildings.

exists, and controls which sense the presence of people, will increasingly appear in lighting control systems.

4. Hours of Use

A key factor identified as highly associated with energy use in buildings in many studies[2,3,15,17,18] is the number of hours per week that a building is operated. It would be expected that a hotel with its 24 hr/day, 7 days/week operation would have a much higher energy use intensity than an office building operated for 9 hr/day, 5 days/week. However, the energy use intensity of a building is not in general directly proportional to its hours of use, due to thermal storage effects.

Analysis of actual energy use in groups of buildings demonstrates the importance of hours of use. The BSRIA Suggested Energy Targets[15] gives the (continuous occupancy) target value of a hotel as approximately double that of single shift offices. In the Wellington CBD studies[2] the building operating hours per week are strongly associated with energy use intensity. In fact our analyses show that the building operating hours (per week) are the most important factor after building size (measured by floor area or number of occupants). The number of hours per week of heating and cooling system operation and building cleaning activities are also significant influences on the energy use intensity of a building.

That the number of hours per week that the cleaners are in a building is an important factor, is the conclusion from several studies of energy used in commercial buildings.[2] With lighting energy use often taking up a considerable fraction of the energy use in commercial buildings, the lighting system hours of use becomes an important energy issue. In many buildings cleaning is done in the evening or at night, hence the lights are likely to be needed. The resultant lighting use can be 10 to 20 hr/week and when added to the, say, 50 hr/week the lights are on for the building occupants, it is clearly a significant additional energy use. Persuading cleaners to turn on lights only when needed, or rescheduling the cleaning to daylight hours, can significantly reduce energy use. Providing the means by which this can be done is an obvious prime requirement of good lighting design.

A common initial reaction to this hours of use question is that it is not an aspect that can be particularly influenced at the design stage. This is not true in general. Very cost-effective energy savings can be made at the design stage, by reducing the hours of use that the major energy-consuming services (usually lights, fans, pumps, heating and cooling systems) will operate. The hours of use of lighting are often very responsive to good initial and detailed design. Good layout of manual light switches and the provision of automatic lighting controls which respond to the level of daylight or the presence of people, can do much to reduce lighting energy use.

There is a possible reason why the building energy use intensity has not been found to be greatly dependent on the centralization of the cooling system (see earlier section). In a climate such as in Wellington, the number of hours of use per year for a cooling plant is on average rather low. This is not the case for a climate such as Washington, D.C. with its relatively long, hot, and humid summers.

Equipment such as pumps and fans frequently operate for long hours and may thus be major energy consumers in a building. Therefore they merit considerable attention at the design stage when the decisions are made that determine the number of hours such equipment will need to be operated. These decisions set limits that even good operational practices could not easily reduce.

There are many situations where the use of relatively inefficient units can actually lead to lower energy use through a major reduction in hours of use. Decentralized air-conditioning units with lower efficiencies than central units are a good example of this. With such decentralized units the occupants will have a greater awareness of their operation, so that they are more likely to be used only when needed.

D. Effects of Users and Occupants

A building may have a number of different tenants. Energy is used in the building systems to maintain comfort both for the occupants who work in the building and for visitors to the building. All the occupants make a direct contribution of heat and moisture to the internal air which tends either to be useful or to require energy for its removal. The occupants may also play a major role in control systems, either directly (for example by opening windows), or indirectly (for example by complaining if conditions are perceived to be uncomfortable).

In practice, a designer's concern ends when the building is occupied but it would be preferable if this concern continued throughout the early life of the building. The designer has much to teach the building users and occupants about how their building should work and has much to learn from them about how the building is used and how it responds.

In the 1960s and early 1970s many buildings were constructed throughout the world where the building occupants had practically no control over the energy-consuming services. In these buildings all the controls were centrally switched either by simple time clocks or by the building manager. Such controls were simple, direct, and inexpensive, but often led to extremely high energy use (see Figures 12 and 13). These 1960s and early 1970s buildings had energy-consuming services features such as constant volume mechanical ventilation, reheat systems, group or whole building switching of lights, and little or no night-setback on thermostats, making them prime candidates for energy conservation retrofits. Recent editions of ASHRAE and CIBS publications show that in such buildings extremely high energy savings can profitably be made by improving the very simple original controls and energy-consuming systems.

At the other extreme there are buildings where the occupants and users have a considerable degree of control over their energy-consuming services and hence over their comfort conditions. Such an approach may appeal to the designer's sense of democracy and allow him or her to sidestep the question of exactly what conditions should be provided: the occupants can decide that! This is certainly a very appealing argument, and the promise of greater occupant satisfaction adds to its attraction.

Giving the building occupants greater control over their energy-consuming services presents several major concerns. These concerns include whether occupants can adequately understand even simple systems, whether the occcupants can play a useful role in buildings with complex systems, and whether occupant control does lead to lower energy use in buildings. Greater individual occupant control is fraught with conflict problems if individual people want different comfort conditions which may lead to, say, simultaneous heating and cooling.

1. Using Occupants to Control Systems

In buildings with spaces that are only intermittently occupied it seems logical to put the local energy-consuming services (lights, heating, cooling, and ventilation) under the control of the occupants as much as possible. It can sometimes be difficult to provide such decentralized control over small areas in a building with, for example, full air conditioning; but for some services, such as lighting, this can usually be achieved. Research at the Building Research Establishment (BRE)[28] has shown that people in continuously occupied offices usually turn on artificial lights if their daylighting becomes inadequate, but are not so good at turning off the artificial lighting when daylighting becomes adequate. Similarly, in naturally ventilated and centrally heated buildings, people will turn on heaters (usually radiators under the windows) when they are too cold, but when they become too hot they are very likely to open the windows and leave the heating on (and waste energy), rather than turn the heating off. This tendency of occupants to respond in an inappropriate manner is not helped by the fact that they are usually not paying directly for the energy wasted. In fact the actual energy cost of inefficient user control practices may not get beyond the organization accountant, so the people squandering the energy receive little or no feedback on the cost of their actions.

Most occupants of nonresidential buildings lack the information, knowledge, and experience to sensibly control even simple decentralized systems. People lack knowledge of how their domestic buildings work and responses in commercial buildings are far less obvious than those in domestic buildings. The Twin Rivers study[12] showed that providing information on energy costs more directly to the home owners leads to considerably reduced energy use. Similarly, for nonresidential buildings, rapid feedback to the building occupants on the energy costs of their actions would be expected to lead to lower energy consumption behavior.

Finally, it may be possible to achieve a compromise between user control and automatic controls that would capitalize on the advantages of both approaches. People are good at turning services on when their environment is uncomfortable and automatic controls are capable of turning devices off when certain comfort (or other) criteria are reached. For example, decentralized light switching can be manual when daylighting is inadequate, and photocells, time-switches, and central controls can turn off the lights when the daylighting becomes adequate or at the end of the working day.

The fact that occupants think that they have some control over their comfort conditions may lead to greater satisfaction, even where such control is very tenuous or even nonexistent. In Wellington, an energy management organization has found that the addition of a dummy thermostat has reduced the number of complaints. Although such an approach should be used only as a last resort, it obviously fulfills a need in some occupants to feel that they have some control over the environment in their working space, even where the nature of the system makes such control impossible. Even with a centralized system, it is often possible to provide thermostats or other devices that allow some sort of voting system to be used, for example, most people may want it hotter or colder at a given time. Such voting systems allow the desires of a majority of building occupants to be reflected in the energy management, in a way that controls operated by individual occupants may not. User controls have the limitation that the individuals who predominantly use them may choose comfort criteria at which other occupants of the space are uncomfortable. Occupants wearing business suits and ties will have different preferred conditions in summer from those people who have changed their clothing styles with the seasons. If the person in the business suit and tie controls the thermostat then it is likely that not only will others be too cold, but also an unnecessary amount of energy will be expended on cooling.

2. Variability of Behavior

Every occupant will have an individual preferred comfort condition and will operate the available controls differently. The studies of optimal comfort conditions referred to in Chapter 3 concentrated on determining the average preferred comfort conditions of a homogeneous group of people, often atypical of the overall population (for example, air-force recruits, university students). Depending on their backgrounds, people's preferred or acceptable comfort conditions vary considerably. An extreme example of these differences would be that between the traditional Japanese approach to domestic heating, which relies mostly on insulating and heating the individual even in very cold winter conditions, and the North American "tradition" of central house heating to 20°C or more. People from such differing backgrounds will obviously not have similar ideas of what are acceptable comfort conditions and hence will set thermostats at different temperatures and operate controls differently.

Even with a very homogeneous population there will be considerable variation in people's energy-related behavior. The Twin Rivers study[12] showed that the effect of individual behavioral differences led to a 2:1 variation in domestic energy use in almost identical houses with occupants who were very similar in background, socioeconomic position, family size, and religion.

Given this variability between occupants, the building designer can only hope to satisfy (or perhaps more accurately, not dissatisfy) most of the occupants most of the time, even if each occupant is given considerable control over their individual environment. If control is given to the users, care should be taken that the control decisions will reflect any majority preference if at all possible. Where appropriate, each user can contribute to a pooled decision via individual thermostats or similar controls, in a system that responds to the desires of the majority of occupants.

3. Effect of Conservation Attitudes

Since the first oil crisis of 1973, considerable publicity has been given to energy matters, at all levels of society in most nations, with special attention being given to oil crises. For example, in New Zealand, "energy" was at first synonymous with electricity. Electricity was temporarily in short supply in the years following 1973 (for reasons unrelated to the oil crisis) so the early thrust of conservation publicity was towards reducing electricity use. More recently in New Zealand the emphasis has shifted towards reducing petrol use, and the emphasis is now slowly shifting towards reducing diesel and jet fuel consumption.

In most nations this changing energy emphasis, coupled with sporadic world crude oil surpluses, has obviously confused the issue of conservation for many people. The popular image of conservation as reducing comfort or convenience, rather than reduction of waste, has not helped the necessary reappraisal of consumer attitudes towards energy use either.

It is fairly obvious that concern for energy efficiency is spreading in most nations, as demonstrated by increasingly efficient transport vehicles and an increased awareness of the potential of solar heating and daylighting for buildings of all types. In existing nonresidential buildings the important question is whether the increasing energy awareness and positive feelings towards energy conservation have led to lower energy use. The building designer is also interested in whether the users and occupants of new buildings are liable to act in a more energy-conscious manner than they did in the past.

The BSRIA study[15] looked at the effect of energy conservation measures in 54 single shift offices and found no marked improvement (reduction in AEUI) for those buildings where investment in energy conservation or energy budgeting had been implemented. The Wellington CBD studies[2] found only a very slight relationship between the question "Was energy conservation practiced?" and AEUI for the 451 buildings where this question was addressed to the building manager.

The success of conservation campaigns in raising awareness, but not causing effective behavioral changes, was demonstrated in a study undertaken by Phillips.[29] It revealed that although people were well disposed toward energy conservation ideas, there was no relationship between what was a high level of perceived need for conservation and corresponding energy-conserving behavior. A social survey carried out by Crossley[30] revealed a similar lack of correlation between beliefs and attitudes on the one hand and energy-conserving behavior on the other.

In the Twin Rivers study[12] it was found that a large proportion of the variation in consumption between houses of similar construction could be accounted for by factors related to the building owners' attitudes to energy conservation, health, and comfort. Owners' attitudes to energy conservation were found to be much less important than those concerned with comfort. Other similar studies have had difficulty showing that energy conservation attitudes lead to measurably lower energy use. Numerous studies have, however, shown that with enhanced feedback on actual energy use or costs, appreciable energy savings are possible.

In both residential and nonresidential buildings then, there is little evidence that energy conservation attitudes alone lead to greatly reduced energy use. This is possibly due to lack of adequate feedback to the building users and occupants on the consequences of their energy-related control decisions. The confusion on what the real national energy crisis is, and a still common belief that it is not a real problem, or is a crisis manufactured by the oil companies to increase profits, does not help either.

Until studies relating changes in energy use to general user and occupant beliefs (of which conservation is only one factor) are available for nonresidential buildings, the importance of conservation attitudes to building energy use must be assumed to be low or at best unproven.

V. RECONCILING PREDICTION AND PRACTICE

The preceding sections of this chapter have provided an overview of both prediction methods (with emphasis on their use at the design stage) and studies of actual building energy use. Unfortunately it has not been possible to provide definitive guidelines for designers attempting to choose between options on the basis of their anticipated annual energy use (cost). Designers will still have to do their own calculations and use their experience to choose between options. For us to present guidelines and pretend these would always lead to energy use reductions would lead designers astray. It should be clear from the previous sections of this chapter that there is still a considerable divergence between what is often predicted to be the energy consequences of a given feature and its observed consequences in actual buildings. To some degree we have set a very high standard of proof, but no more than would be expected in any other scientific discipline.

The following sections will summarize the points made in previous sections of this chapter, suggest reasons for the divergence between prediction and practice, and examine anticipated future developments.

A. Reasons for Divergence

We have already discussed the lack of success in attempts to prove, via energy use studies, that any significant correlation exists between building energy use and the features that are conventionally assumed to have a significant effect on it. In the absence of actual observed correlations the following three hypotheses must be considered:

1.   The prediction methods used are neglecting vital factors.
2.   Some of the factors actually considered are incorrectly modeled.
3.   The input and output of these methods are misinterpreted.

It is our contention that elements of all three hypotheses are present.

With few exceptions the climate, design, end-uses, and operation of a building are unique. Most buildings are one-off designs and normal production tolerances of components and differences in construction lead to significant variations between even nominally similar buildings.[12] The effect of such variations is difficult to quantify. The discovery, for example, that natural infiltration rates vary considerably between similar buildings gives some idea of the difficulties involved.

There are many people intimately involved in a building, ranging from the designer, through the building users and occupants, to the cleaners and maintenance personnel. To some degree all these people add further cumulative uncertainty to energy use predictions. Such imprecisely known interactions between people and systems must lead to a considerable "fuzziness" in the relationship between the factors that can be predicted at the design stage and the resultant actual energy consumption of a building. Some of this uncertainty should be quantifiable in future prediction models. However, a considerable degree of imprecision will remain, some of which is the inevitable result of the assumptions that must be made if a prediction method is to be useful.

Present prediction methods are rather uneven in their treatment of the different modeling areas, some being very detailed, others crude in the extreme (usually due to a lack of knowledge in the area). Daylighting and air infiltration are two areas where few general purpose prediction methods yet incorporate the best models available. It would seem that few of the people intimately involved in formulating prediction methods have looked at their accuracy requirements dispassionately; what is the point of using very complex and highly accurate calculation routines if the input data is only approximate?

The validation of most prediction methods has been neglected in the apparent rush to produce better, faster, and more comprehensive versions of the computer packages. Such validation that is readily available is often of a few individual buildings and usually shows reasonable accuracy once the fuzzy factors can be determined accurately, more accurately than is possible at the design stage. Going to an existing building to measure the fuzzy input data variables is a form of backwards validation commonly used. Such backwards validation is useful in isolating program code errors and shows that programs are useful for retrofit studies of existing buildings.

What is of most interest to the building designer, however, is whether such prediction methods can be accurately used in forward looking design, a question that most of the available literature on program validation does not address. It is usually assumed that whatever the limitations of the prediction method, the relative ranking of the energy use of different building options is valid. This assumption should be treated very carefully indeed, given the sensitivity of actual energy consumption to minor factors (such as time-clock settings).

Simpler prediction models that are cheaper and easier to use would seem to offer considerable advantages. Such models would not be able or expected to predict the energy use of every conceivable building, but specialist programs would be available for unconventional designs. Simpler models would encourage the program user to explore more options and examine the sensitivity of the predicted energy use to uncertainty in input data values or commonly made assumptions. Perhaps most importantly, simpler models would reduce the very human tendency to regard the results as highly accurate predictions of energy use or energy cost (the tendency to predict the energy cost of a building in x years as $489,651 instead of a more realistic $500,000).

Energy prediction methods certainly have their place, as their widespread use by consulting engineers demonstrates, but there seems to be a lack of appreciation of the compromises and assumptions that must be made in any useful prediction method. The lack of comprehensive published validation makes it difficult to gauge the limitations of these prediction methods. Therefore a healthy scepticism should be used when interpreting predicted energy use values.

B. Energy Use in Practice

There is a general scarcity of systematic studies for ordinary nonresidential buildings which examine the relationships between actual building energy use and building features. The results of those studies that the authors have examined[15,17,18,23,26] are briefly summarized in this section.

All the studies have shown there is a great deal of variation in building energy use intensity, even within buildings housing similar activities and subjected to the same macroclimate. This observed variability extends into subgroups of buildings; fully air-conditioned buildings, for example, have a large AEUI range.[2] While it is possible to find differences between the average AEUIs of, for example, mechanically ventilated and naturally ventilated buildings, these differences are usually found not to be significant. In fact, once the energy use has been divided by a measure of building size, only hours of use and measures of building complexity can split a group of buildings into statistically significant subgroups.

Analytical techniques such as multiple regression analysis can be used to identify, in greater detail, the building parameters associated with energy use. The actual factors identified as being highly associated with energy use differ somewhat between studies, but a few general trends have been observed. The hours of operation of a building and of its heating and cooling systems, as well as those of the cleaners, have been found to be associated with higher energy use intensities. Increased centralization of services has also been found to be associated with higher energy use intensities.[2]

Increasing climatic severity, as measured by heating degree-day totals, is accompanied by higher average energy use intensities, but the linkage is perhaps weaker than is commonly assumed (see Table 1). Certainly the energy use intensities of commercial buildings seem to be less directly related to climatic severity than is the case with noncommercial buildings (see Figure 4 and Table 1). This reduced dependence on climatic severity tends to confirm the concept of commercial buildings usually being slow to respond to climatic changes and self-heating (from internal gains) to a considerable degree. The effect of climatic variation on the energy use of a given group of buildings can be established within a given year (see Figure 3) but is more difficult to establish between years as there are the confusing effects of responses to the energy crises as well as changes in the building stock to be considered. However, if climatic dependence can be observed for different locations and within a given year for a group of buildings, then it is likely that, everything else being equal, energy use will change from year to year in response to climatic changes.

In studies of energy use in groups of nonresidential buildings the architectural factors of building form and fabric have not been found to make a noticeable difference to the energy use intensity. This suggests that the potential contribution of architectural factors can only be realized in buildings with well-integrated designs that are carefully constructed and operated. Examples exist where this has been achieved in practice for single buildings. However, designs in which architectural factors are claimed to contribute to lower energy use should be treated with caution until actual energy use data is available and the resulting occupant conditions can be evaluated.

The building users and occupants, where they play an active role in controlling building services, are undoubtedly associated with a significant amount of the observed variability in energy use between similar buildings. However, the precise amount has not yet been quantified. In buildings where the users and occupants play little part in the control of building services, the observed variability in energy use is probably also due to factors related to people, in this case those involved in the commissioning, balancing, maintenance, and operation of control systems.

The influence of the occupants' attitudes to energy conservation are also difficult to quantify. The results of the few systematic surveys of commercial buildings (and of some domestic building studies) suggest that attitudes alone have very little influence on energy use behavior. In the absence of strong incentives to conserve and good direct feedback on the energy use consequences of control decisions, the occupants are unlikely to match their enthusiasm for conservation with effective conserving action.

C. Closing the Feedback Loop

In the system of building construction presently practiced throughout the industrialized world there is a general lack of feedback, at all levels, on the energy use consequences of peoples' actions.

Building designers are unlikely to examine the energy use of a building once it has been properly commissioned (if in fact it ever is) and to compare this actual performance with the predicted performance made at the design stage. The designer wishing to check the calculated energy performance of a new building, against that of similar existing buildings, is hampered by the paucity of information on the energy use characteristics of existing buildings.

The lack of feedback is one of the more pressing problems facing building designers interested in the energy performance of their designs. Many of the impediments to effective feedback are institutional; for example, professional fee structures make no provision for an evaluation of the energy performance of the building (let alone of any other aspects of the performance of the building).

It is regretted that more positive guidance could not be given on the lowest energy use systems that building designers could or should use. We feel that there is insufficient evidence available to support definitive statements such as, say, natural ventilation should be used wherever possible. We can only hope that the contents of this chapter will encourage designers to examine critically all the options available. We also hope that designers will carry out many calculations and simulations, but only after looking at similar existing buildings to make sure that they are concentrating on the most important areas relating to energy consumption.

REFERENCES

1. **Jencks, C.,***The Language of Post-Modern Architecture,* Academy Editions, London, 1981.

2. **Baird, G., Donn, M. R., and Pool, F.,** Energy Demand in the Wellington Central Business District — Final Report, Publ. No. 77, New Zealand Energy Research and Development Committee, Auckland, New Zealand, 1982.

3. **McCarthy, P. M., Patel, R. F., and Karpay, B.,** Empirical and Simulation Analyses of Energy Use in Commercial Buildings, HIT - 664-2, Hittman Associates, Columbia, Md., February 1977.

4. **Kusuda, T.,** Review of Current Calculation Procedures for Building Energy Analysis, NBSIR-80-2068, National Engineering Lab., National Bureau of Standards, Washington, D.C., July 1980.

5. **Pool, F.,** Computer Simulation of Building Energy Requirements, M.E. Project Report, Department of Mechanical Engineering, University of Canterbury, Christchurch, New Zealand, 1977.

6. **Mitalas, G. P.,** Net Annual Heat Loss Factor Method for Estimating Heat Requirements of Buildings, in Building Research Note 117, Division of Building Research, National Research Council of Canada, November 1976.

7. **Hittle, D. C.,** The Building Loads Analysis and System Thermodynamics Program, Vol. 1, U.S. Army Construction Engineering Research Laboratory, Champaign, Ill., December 1977.

8. **Curtis, R. B.,** The Theoretical Basis of the DOE 2 Building Energy-Use Analysis Program, LBL-12300, Energy and Environment Division, Lawrence Berkeley Laboratory, University of California, April 1981.

9. **De Winter, F.,** Solar radiation — a national disgrace, *Solar Age,* January 1982.

10. U.S. Department of Energy, Comparison of Load Determination Methodologies for Building Energy Analysis Programs, Final Report for International Energy Agency, January 1981.

11. **Kusuda, T.,** A comparison of energy calculation procedures, *ASHRAE J.,* August 1981.

12. *Energy and Buildings,* 1, 3, Elsevier Sequoia S.A., Lausanne, Switzerland, April 1978.

13. **Elder, H.,** Energy Conservation in Hospitals, Proc. Tech. Groups (Building Services), New Zealand Institution of Engineers, Wellington, New Zealand, 1983.

14. **Wortman, D., O'Doherty, B., and Judkoff, R.,** An Overview of Validation Procedures for Building Energy Analysis Smulation Codes, NTIS Microfilm SERI/TP 721-1133, Solar Energy Research Institute, Golden, Colo., March 1981.

15. **Stewart, L. J. and Colreavy, J. P.,** Summary of Report on Collection and Analysis of Data for the Assessment of Building Targets, Building Services Research and Information Association, Old Bracknell Lane, Bracknell, Berkshire, England, 1980.

16. **Ternoey, S. E.,** Don't trust your instincts: big building design defies intuition, *Solar Age,* October 1981.

17. **Beca Carter Hollings and Ferner, and Shaw, R. A.,** Greater Auckland Commercial Sector Energy Analysis, Rep. No. 45, New Zealand Energy Research and Development Committee, Auckland, New Zealand, May 1979.

18. Syska and Hennessy and Tishman Research Corporation, Energy Conservation in Existing Office Buildings, Phase 1 Report, Department of Energy, Washington, D.C., June 1977.

19. **Hill, E. H. et al.,** Performance of the Norris Cotton Federal Office Building for the First Three Years of Operation, Building Science Series, National Engineering Lab., National Bureau of Standards, Washington, D.C., August 1981.

20. **Donn, M. R. and Pool, F.,** 1981 Annual Building Energy Use Survey for the Wellington Central Business District, Tech. Publ. No. 17, Ministry of Energy, Wellington, New Zealand, July 1982.

21. **Stein, R. G.,** Research, Design, Construction and Evaluation of a Low Energy School Research Phase 1, Interim Report, National Science Foundation, Washington, D.C., August 15, 1974.

22. EFL, Energy Study — sponsored by the Fairfax County School Board with financial support from the EFL Inc. to determine the potential for energy conservation in the Public Schools, Education Facilities Laboratories, New York.

23. **Livesey, P. M.,** National savings, *J. CIBS,* 3, 8, 51, 1981.

24. **Baird, G. et al.,** Performance evaluation of air conditioning and related energy consuming systems, by computer, in Int. Congr. Air Cond. and Computer, Milan, Italy, March 1982.

25. **Banham, R.,** *The Architecture of the Well-Tempered Environment,* The Architectural Press, London, 1969.

26. Enviro-Management and Research, Evaluation of Building Characteristics Relative to Energy Consumption in Office Buildings, Washington, D.C., September 1975.

27. **Setly, B. S. V.,** The nation's most energy efficient office building, *ASHRAE J.,* November 1979.

28. **Hunt, D. R. G.,** The use of artificial lighting in relation to daylight levels and occupancy, *Build. Environ.,* 14, 21, 1981.

29. **Phillips, P.,** Household Energy Consumption Attitudes, Rep. No. 10, New Zealand Energy Research and Development Committee, Auckland, New Zealand, 1976.

30. **Crossley, D. J.,** Social Factors Affecting Energy Use and Conservation in the Home, AES Working Paper 1/80, School of Australian Environmental Studies, Griffith University, Australia, 1980.

Chapter 7

USER CONCERNS — ENERGY MANAGEMENT AND ANALYSIS

I. INTRODUCTION

The concerns of management, in relation to the energy performance of buildings, will be examined in this chapter. Particular emphasis will be placed on methods for initiating a conservation program within a group of buildings, using a minimum of information, resources, and analysis.

The thrust of this chapter is to examine how a selection of existing energy management or conservation programs function and to describe some simple techniques for guiding the critical early stages of such programs. Technical details of the maintenance and upgrading of equipment are not covered here as they have been dealt with extensively elsewhere.[1]

In the first section some of the impediments and opportunities facing those engaged in building energy management are discussed. The second section summarizes the approach taken by a range of government agencies in the U.S., the U.K., Canada, and New Zealand to the energy management of large numbers of buildings. In the third section, two building selection techniques are described, for energy conservation programs covering large groups of buildings. Of the two methods presented, the first, Pareto analysis, highlights the significance of the large consumers in a group of buildings; the second method describes the use of energy indexes for interbuilding comparisons and focuses on the assessment of the conservation potential of individual buildings within a particular group. The fourth section deals specifically with two techniques that can be applied to individual buildings. One is the use of load profiles to establish when and for what purpose energy is consumed in buildings; the other is the use of performance lines to monitor thermal environmental control systems. In the final section, a strategic framework for building energy management is proposed.

II. IMPEDIMENTS AND OPPORTUNITIES

A. Exogenous Factors

The building manager operates a system that has predominantly been designed by others, in particular the building owner and designer as described in previous chapters. This results in physical and operational constraints on the manager. These constraints include the building location, orientation and configuration, its materials and contruction, and its energy-consuming systems and controls. The building manager normally must accept these constraints, at least at first.

The first concern of management is to establish where energy is being used in the building, when it is being used, and how much is being used. Low cost operational improvements can then be attempted, followed by higher cost measures such as the recommissioning of energy-consuming systems. Once these have been done, further analysis of the building as an energy-processing system may show major modifications to the building fabric and the energy-consuming services to be justified. An energy consultant is best engaged at this point to ensure that the most cost-effective approach is adopted.

Besides the constraints inherent in operating an existing building, the energy manager faces the practical problem of establishing the details of the actual building design and the intended system operation. As-built drawings may not be available and the original design drawings are unlikely to provide an accurate record of the energy-consuming systems. If the initial system balancing and commissioning was inadequate, if maintenance has been on an ad hoc basis, or if repairs are carried out by those unfamiliar with the design, then the actual mode of operation may be markedly different from that intended. It is not uncommon for example to find fans running in reverse, heating and cooling coils working in opposition, inoperable dampers, and so on. Many of these defects can be detected by visual inspection. However, a systematic appraisal of energy performance is hamstrung unless accurate information is available. Drawings of the installed design, plus specifications and operating manuals supplied by the original building design professionals and equipment manufacturers, are especially useful.

The changes in energy prices and availability that affect all consumers are another major influence on the energy manager. Government policies, together with those of energy suppliers, attempt to match supply with demand (and vice versa). This often results in energy prices which are subject to frequent and unpredictable change. Besides energy prices, there may be incentives, such as tax write-offs and low interest loans, to encourage the use of certain fuels or the installation of energy-conserving equipment. To effectively control energy consumption, energy managers must have adequate information about the energy performance of their buildings, in order to assess the impact of price changes and the value of incentives.

The other major constraint on the operation of buildings is the prevailing set of environmental standards (see Chapter 3) to which the building manager must adhere. These are continually being adapted in recognition of the scarcity value of energy. Thermal comfort, ventilation, and lighting requirements have all been reconsidered since energy efficiency became a paramount design and operation criterion. To take effective advantage of these new attitudes the energy manager needs to be aware, not only of these new standards, but also of the standards originally used in designing the services of a given building.

B. Impediments to Energy Management

In most sectors, a fundamental distinction between the production and the consumption of energy is that a small number of large production facilities supply a large number of relatively small consumers. It is the diffuse nature of this consumption which makes energy demand reduction a complex problem involving the interplay of technical, economic, institutional, and psychological forces. The buildings sector follows this general trend.

Widespread examples of inefficient building services design and control exist and offer scope for energy cost savings. There is every indication that major cost-effective savings can be achieved, for example, by improved maintenance and simple adjustments. Unfortunately, a variety of technical, institutional, and information impediments tend to prevent the full energy cost savings potential from being realized.

Prior to comparing and evaluating different approaches to energy conservation in buildings, it is useful to outline the impediments and identify features which help an energy management program. This provides a background against which the various approaches may be assessed.

1. Technical Impediments

Buildings have a number of characteristics which make their energy performance difficult to monitor and manage effectively. Frequently, one of the first impediments relates to the building services design. For example, existing electrical wiring layouts and central heating systems usually prevent individual tenants in the same building being separately metered for all their energy use. Consequently, direct cost feedback cannot be given, and apportioning of energy costs is often done on the basis of floor area occupied.

Heating, ventilating, and air-conditioning (HVAC) systems are normally designed to operate without direct control by the building occupants. Equipment such as pumps and fans must often operate during periods when the building is unoccupied. All but the simplest systems are usually automatically controlled. Consequently, it is often not possible for the user to detect inefficient services operation. Plant rooms are isolated and occupants are rarely aware of what the equipment and its controls are designed to do. Detection of gross operational inefficiency usually occurs only when there is a perceptible decline in environmental conditions or the energy cost becomes unacceptably high.

As discussed in Chapter 2, yardsticks of energy performance are not familiar to building users, nor are they necessarily applicable to comparisons of energy consumption at different times. Unlike their counterparts controlling industrial processes, building managers cannot normalize the energy consumption of their building by some readily available factor, such as tonne of output, in order to give some comparable measure of energy performance. Because of a lack of published data, the normalized measures that do exist, such as AEUI, are often limited in their utility.

Reliable comparisons of building energy performance are best made when the components of total consumption which are affected by different factors, e.g., electricity for lighting, gas for heating, can be separated. It is also desirable if each component of consumption can be normalized with respect to all the factors which influence it (e.g., temperature, solar radiation, operating hours). However, such comparisons are much more difficult than, for example, using a liter per 100 km (or mi/gal) measure for the fuel consumed by motor vehicles.

2. Institutional Impediments

Institutional impediments take many forms, but the following examples illustrate some of the major types.

Our first example involves leasing agreements. These usually determine whether individual tenants purchase energy directly from the supply company or pay an apportioned share of the total building energy consumption indirectly in their rent. Leases which include an allowance for energy costs in the rent are undesirable from an energy management standpoint, because they permit owners to pass on costs directly to tenants, by way of cost escalation clauses. Since building owners do not bear these costs, there is no direct incentive for them to reduce energy consumption. The limiting factor is when rent charges become uncompetitive.

A second kind of institutional impediment often occurs in large buildings, where energy is a common property resource. A type of "tragedy of the commons" situation exists[2] where anyone can benefit from high consumption of energy, while the costs are spread over all the occupants (indirectly as increased rent). In this case, the individual who conserves does so for very little benefit, since any savings accrue to all occupants (through lower rent).

A third example of this type of impediment is the lack of technical expertise among most building owners and tenants (in comparison, say, with the industrial sector). Rarely are there sufficient in-house skills to carry out the balancing and maintenance required for energy-efficient building operation. A drastic event, like a plant failure or a major energy price increase, is usually required before consultants are employed. In addition, the fact that most building owners specialize in nontechnical fields will affect their investment decisions; capital expenditure for energy conservation is unlikely to be an area of expertise.

3. Information Impediments

Lack of knowledge about energy-consuming systems often impedes attempts to reduce consumption. For example, plant performance characteristics are rarely known outside the building services field. Most car owners are aware of the fuel consumption/performance tradeoffs between energy and engine capacity, automatic and manual transmission, etc. Building owners on the other hand, rarely know much about the energy consumption/ performance tradeoffs of dual-duct, reheat, or variable-volume air handling systems. Again, motorists know that the fuel consumption of different models can vary by up to a factor of three, but building owners are almost certainly unaware that the AEUI of a commercial building can vary by a factor of ten.

Unlike the homeowner, occupants of commercial or institutional buildings usually have no idea of the energy costs of operating their buildings. The feedback provided by monthly or bimonthly energy bills seldom leaves the accounts department.

C. Energy Management Opportunities

While the preceding discussion of impediments indicates some of the difficulties facing building energy management, in practice, many useful conservation opportunities still exist. The examples that follow highlight some of the characteristics of buildings and groups of buildings which may be used to advantage in an energy management program. Large consumers, although a low percentage of the number of buildings in a group, tend to dominate that group's energy consumption. In the Wellington Central Business District, for example, the top 10% of consumers accounted for 70% of the CBD consumption.[3] Major savings in the CBD energy consumption could most readily be made by reducing the energy use of a few of the large energy consumers. An average reduction of only 14% in the top 10% would reduce the energy consumption of the whole CBD by 10%.

Building supervisors are often employed in larger buildings where the results of conservation are likely to be the most significant. In-service training of these personnel to improve their proficiency in operating plant, identifying malfunctions, etc., is likely to be the most effective educational strategy in a management program.

In commercial buildings, fuels and their metering arrangements often provide a distinctive split with respect to end-uses. Fossil fuels tend to be used for space heating and minor domestic hot water requirements. Electricity supply authorities may have separate tariffs covering different end-uses, for example "power" and "lighting". These distinctions tend to aid the analysis of energy consumption within commercial sector buildings.

Examination of an existing building, with established use patterns, enables a much more precise definition to be given of the environmental requirements of various activities than was possible at the design stage. This means that any energy-consuming systems involved in meeting these environmental requirements can be adjusted to suit, and that the activities themselves can be scheduled in such a way as to minimize operation of the systems.

III. SELECTED ENERGY MANAGEMENT PROGRAMS

The following section describes the main elements of a few building energy management programs. It provides an insight into the way various agencies have managed the energy performance of their buildings.

The administrative, financial, and educational aspects of energy management are vital components of any successful program. A structure which uses the information obtained from energy monitoring and analysis techniques is essential to good management. Energy management programs will founder unless responsibilities are clearly defined. In addition, the criteria for funding conservation measures must be established so that scarce financial resources are used to the best advantage for the particular organization involved. In a successful energy management program, the impediments discussed in the previous section must also be overcome.

There must be very few organizations using buildings who have not already contemplated an energy conservation program for their own building stock. What follows is a sampling of the approaches taken by a few agencies in four nations. Many of these programs are in their initial phases, so the sample contains many ideas worth attempting, but less confirmation of actual achievements. The programs selected illustrate a range of organizations, with differing internal administrative structures, numbers, and character of buildings, geographical coverage, and energy expenditures. All of them are governmental organizations (from whom data tends to be more readily available). The following agencies are examined:

U.S.
- The General Services Administration (Federal Government)
- New York City Municipal Government
- City of New York Board of Education

U.K.
- The Property Services Agency (National Government)
- Strathclyde Regional Authority
- Cheshire County Council

Canada
- Ontario Provincial Government
- Municipal Initiatives and Programs
- City of Toronto Board of Education

New Zealand
- Ministry of Works and Development

A. U.S.

Energy conservation in buildings rates high in the order of conservation priorities of the U.S. Programs abound at federal, state, county, and municipal levels.

1. The General Services Administration (GSA)

The GSA is responsible for the approximately 5000 buildings owned by the Federal Government of the U.S., together with a further 5000 or so leased premises. Responsibility for the operation of these buildings rests with the Public Buildings Service (the PBS is one of six major departments of the GSA). The Energy Conservation Division of the Office of Building Management (one of the four offices of the PBS) carries out the energy conservation programs. The country is split up into 11 regions for administrative purposes.

Taking 1975 as the base year, a 10-year energy conservation plan[4] with the following target reductions is being pursued:

Existing owned buildings	20%
New owned buildings	45%
Overall oil consumption	30%
Existing leased buildings (1980—1985)	5%

As far as energy monitoring of owned buildings is concerned, monthly fuel use or delivery data is collected at regional level (from suppliers' invoices) and input to the GSA computer system. Approximately 570 buildings, representing 80% of the floor space owned by GSA, are being monitored. The data is analyzed by staff of the Energy Conservation Division.

In the first 3 years of the program there was very little change in energy use in existing owned buildings, while for new buildings a 30% decrease was achieved. Overall oil consumption decreased by 6%.[5] As well as energy monitoring, the Energy Conservation Division is involved in operation and maintenance, retrofit, and employee awareness programs. More staff were appointed at the regional level — from 3 to 15 per region, depending on size — to carry out this work.

2. New York City

New York City is the largest municipal energy consumer in the U.S. The Office of Energy Conservation, of the Department of General Services, functions as a central monitoring and guidance center for the conservation program of the city. Nearly 4000 buildings are involved, with annual energy costs of U.S. $400 million in 1980. The Office of Energy Conservation (OEC) has been in operation since 1976 and is responsible for four main activities:[6]

- Paying all utility bills — thus combining the twin functions of financial and energy monitoring
- Monitoring energy consumption — this is carried out via energy liaison officers, one for each of the 58 city departments; each department is responsible for the development of its own energy conservation program; the OEC has produced a manual[7] to assist the liaison officers in this task
- Implementing a U.S. $100 million retrofit program over a 4-year period and covering 200 city properties; 19 firms of consulting engineers and architects were involved in the energy auditing and design work associated with this activity
- A public relations and information campaign, designed to motivate building managers and occupants to support energy conservation practices in the buildings they use

3. City of New York Board of Education

The Office of Engineering Support Services of the Division of School Buildings has been active in the field of energy conservation for many years. However, as the largest consumer of energy of all the departments of the city, it continues to strive for increased efficiency. A considerable amount of retrofit work has been carried out, aided by federal and state grants in most cases.

Considerable emphasis is placed on assisting school custodians to examine the operation of the buildings under their care from an energy conservation point of view. A manual[8] has been prepared and the incentive to use it is strong, as the custodians' annual promotion rating includes energy conservation performance in its calculation. Energy monitoring involves each school sending in to the department a "Monthly Fuel and Utility Report" which is collated manually and summarized on an annual basis.

B. U.K.

Energy conservation in buildings has received a considerable amount of attention at both national and local level in the U.K., as will be evident from the following cases.

1. The Property Services Agency

The Property Services Agency (PSA) of the Department of the Environment has responsibility for all Government buildings in the U.K. Energy costs for 1980 to 1981 were about £220 M. By any standard, the results of the PSA energy management and conservation program are remarkable.[9] Overall savings of 35% were achieved over a 6-year period (1972 to 1973 was the base year). The three main aspects addressed, as part of their conservation program, were the upgrading of heating controls, the training of operating and maintenance staff, and the giving of advice to building users and operators.

Subsequently, a new target, aiming for a further 12% saving is well on the way to being attained.[10] One of the principal means of achieving this new target was the identification of the 20% of properties which consume around 80% of the energy in government buildings. This identification and ranking, based on annual consumption data, was carried out by the PSA Energy Conservation Group. This group comprises four sections dealing with computer applications, control engineering, energy conservation (existing estate), and energy planning (new designs).

A small multidisciplinary team of energy conservation officers has been set up in each PSA region. The Scottish region, for example, with annual energy costs of some £7 M, has a headquarters team of 4 professionally qualified and experienced engineers and architects, together with 17 engineering plant operators (a new class of multiskilled craftsmen) distributed throughout the region.

These regional headquarters teams are responsible for overall monitoring of the top 20% of consumers and for allotting target consumption figures to individual buildings. Area Works Offices in each Region provide the link to the District Works Offices which have the primary responsibility for the operation of the buildings (typically around 200) under their care. At the building level, the Accommodation Officer has the responsibility for submitting quarterly reports of energy consumption. It is this data that provides the basis for sound energy management decisions, such as the need to investigate the operation in more detail of a particular building or the requirement for a consultant's report in order to assess energy conservation priorities.

2. Strathclyde Regional Authority

Strathclyde, the largest regional authority in Scotland, cares for 500 buildings, on which it spent £25 M on energy in 1979, and has 2000 building caretakers on its staff.[11] The four-person Energy Conservation Unit at the Department of Architectural and Related Services of the region was set up in 1978. It is responsible for energy price negotiations, a program of caretaker and building inspector training, the direction of three energy conservation field teams, and the instigation of an energy-monitoring system.

The energy conservation field teams have concentrated on checking the efficiency and operation of boiler plants, producing considerable savings as a result. The budget of the unit was not sufficient to fund major retrofit work; this is to be financed by the department which is using the building.

3. Cheshire County Council

The energy bill of this council amounts to some £7 M per annum (1979/1980 figures).[12] An Energy Conservation Unit, set up in 1975, has developed an energy-monitoring system by which consumption data is sent to it from each of approximately 1200 individual buildings, using a prepaid postcard system. Approximately £0.25 M has been spent on installing meters (for oil and water especially) and an estimated £100,000 (equivalent to three person-years) on the development of an energy-monitoring computer program.

Cumulative savings of more than £1 M were attained over a 3-year period. The target for the council is for an annual saving of £1.3 M (1977 figures) and there is every indication that this will be achieved.

C. Canada

The Canadians too have placed a high priority on energy conservation in buildings. Public Works Canada, the Division of Building Research of the National Research Council and Energy Mines and Resources Canada, are active at the national level. The following notes focus on provincial, regional, and municipal level activities.

1. Province of Ontario

As a net importer of energy, the Province of Ontario is in a relatively vulnerable position with respect to energy supplies. This was evident, for example, in the 1976 to 1979 energy research budgets of the provincial government, most of which appeared to be focused on conservation.[13] Many individual ministries (e.g., Energy, Health, Education, Housing) have active programs related to the reduction of energy use in the buildings they themselves occupy.

The Ministry of Education, for example, is involved in the collation of energy use data from schools. The Ministry of Health, with 240 hospitals and annual energy costs amounting to CDN $100 million requires quarterly reports of energy consumption and costs. A capital grant program has been instituted for energy conservation measures with predicted payback times of 3 years or less. The Ministry of Energy, apart from its overview of the other government departments, has already achieved a target reduction of 15% in energy use in its own 3.5 million square meters (38 million ft²) of accomodation, within 4 years of starting.[14]

2. Municipal Initiatives and Programs

Arguably a significant watershed was the "Cities Energy Conference" held in Toronto in early 1980, together with the survey preceding it.[15,16] This is not the place to attempt a summary of the findings of such a major exercise. What became clear, however, was that the municipalities themselves needed to take the initiative and could not await federal or provincial recognition of their particular problems. The reports produced for the conference describe the broad spectrum of municipal energy conservation measures, both in-house and community directed, which were undertaken in 20 Canadian cities.

The city of Toronto, for example, carried out an audit of the energy consumed by the various municipal operations which in 1979 were estimated to be costing CDN $30 million per year, or about CDN $50 for each resident of the metropolitan area,[17] while the neighboring city of Mississauga had an energy consumption reporting system already in place for its municipal buildings.

3. City of Toronto Board of Education

In many parts of the world, school buildings have tended to be a readily identifiable group for energy conservation. The 155 schools in Toronto are no exception. As with other groups of schools, caretaker training was seen to be a key issue, and a 3-week training program was introduced for these personnel. In addition, remote monitoring of energy-consuming plant was implemented, using a central computer located at the headquarters of the Board. As a direct result of conservation measures carried out between 1975 and 1978, cost savings of almost 13% (or CDN $840,000) were achieved in 1978, in comparison with the 1974 base year.[18]

D. New Zealand

The New Zealand Government, through its Ministry of Works and Development (MWD), has established a Building Energy Conservation Advisory Service[19] for government buildings. The MWD is responsible for the design, contract arrangement, construction supervision, and maintenance of the buildings occupied by most government departments. It is thus the logical agency to provide an energy conservation service.

The guiding principle of this service is that the occupants of government owned or leased premises, together with the government accommodation agencies, have a joint responsibility for the efficient use of energy in the buildings they control and occupy. The MWD has established a multidisciplinary building energy conservation advisory team in each of its seven districts. In addition, individual government departments appoint an energy manager (assisted where necessary by energy wardens) for each building under their control.

The multidisciplinary teams each consist of a mechanical and an electrical engineer together with an architect. The team's duties are to analyze the energy use of a building, conduct energy audits as resources permit, and recommend ways of reducing energy consumption. The teams are also responsible for the organization of training programs for the building energy managers and wardens.

The building energy managers represent the interests of the building occupants and provide the link to the MWD team. They are responsible for the monitoring and reporting of building energy use to the head office of the department and the MWD, for the dissemination of information on energy conservation to the building occupants and for the implementation of measures to reduce energy use.

Within a short time of the commencement of the program, worthwhile savings in the annual energy bill of the government building sector were achieved. Adjustment of heating controls has resulted in savings of 25 to 40% in a number of buildings. More energy-conscious use of lighting systems has produced savings of 10 to 16%. The MWD teams have repeatedly found building services installations with faults of the following types:

- Incorrect installation
- Never commissioned
- Improperly commissioned
- Recommissioning required

The MWD Building Energy Conservation Advisory Service will continue to develop in the foreseeable future. It is planned that the energy consumption data be categorized by end-use and that realistic energy performance criteria be set for government buildings. It is also intended that the service be made available to local authorities, as well as to national government agencies.

As far as the evaluation of specific conservation proposals is concerned, "the principal economic guideline to all officers concerned with energy conservation in government buildings is that the capital, operating, and maintenance costs of any energy conservation scheme must be recoverable from the national value of the energy saved."[20]

E. Energy Management Program Guidelines

The previous section illustrated that there is no standard approach to energy management. Programs must be tailored to the requirements of the organization concerned, taking full account of all its needs and resources. However, some general principles and guidelines have emerged and these will now be outlined.

1. Program Administration

The administrative set-up will depend on the number of buildings involved and their geographical spread. The use of direct computer-based energy management control seemed a logical option for the city of Toronto Board of Education, with its relatively small numbers of buildings and fairly compact geographical area. However, this is not an option that was attempted by any of the larger authorities reviewed, most of which grafted energy administration onto their existing organizational structures.

2. Financial Considerations

At least three aspects should be considered here:

- An energy management program often gives an overview of individual fuel costs and tariffs, thus providing a suitable basis for negotiation with energy suppliers. The Strathclyde Regional Authority was successful in this area.
- The channeling of fuel bills through an energy conservation administration, as practiced in New York City, is a convenient and logical way of expediting the monitoring of energy use and costs.
- Sources of finance often have to be found within the organization to fund any major conservation effort, and the beneficiary of any savings must be clearly identified, perhaps the most vital considerations of all.

3. Energy Monitoring

This is an essential part of any worthwhile program. It may be surprising to learn that many organizations do not have a complete and reliable list of their buildings; the compilation of such an inventory is often the first task of any energy management program.

It will often be necessary to install additional energy meters — Cheshire County Council found this to be a worthwhile exercise — in order to obtain the relevant information. The resulting data may be collected and processed manually or by computer, locally or centrally, according to preference. It is important that the data be reliable and the frequency of collection (and the resulting feedback) appropriate to energy conservation priorities. The Property Services Agency (PSA), for example, specifies a higher frequency of energy data collection for the top 20% of consumers than for the rest.

4. Training and Education

Training and education have been given a high priority by most authorities. Training programs for caretakers, plant operation, and maintenance staff and others directly involved in running the energy systems of buildings have been instituted, while more general conservation campaigns have been directed towards the building user.

5. Analytical Tools

A good range of analytical tools is necessary in order to translate the monitored data into information which can be used to make energy management decisions. The U.S. GSA perceived the need for information to be produced in this form, the U.K. PSA tested various analytical methods.

The remainder of this chapter will concentrate on the development and demonstration of analytical methods, the results of which may be employed as a basis for decision making and which, taken together, form a coherent strategic framework for building energy management.

IV. MANAGEMENT METHODS FOR GROUPS OF BUILDINGS

Once the initial level of resource commitment to a program has been established and responsibilities have been assigned, there is a need to decide how much information is required before making investment decisions. The following two sections describe methods and techniques of building energy management which may be employed at the early stages of a program, using a minimum amount of information. Prior to describing management techniques applicable to individual buildings (see next section) we shall examine methods applicable to groups of buildings.

For property managers with a large number of buildings under their care, one of the most basic problems is identifying those individual buildings which are most likely to yield cost-effective energy savings. In order to do this, techniques for ranking energy use, often utilizing existing records, are used.

One such technique is Pareto analysis which can be used to rank the buildings in a group, according to their energy consumption. Such an analysis enables a target subgroup of buildings to be identified for a more intensive conservation program. The method will be described in the first half of this section.

Depending on the size of this target subgroup of buildings, it would be useful to have a method for choosing those to be examined first. In the second half of this section, a simple index, which can be calculated with a minimum amount of physical and energy data about the buildings, is described. This index can be used to rank buildings according to their energy conservation potential.

A. Pareto Analysis

Pareto analysis provides the first stage of an analytical approach to the problem of allocating limited resources to achieve maximum results. The method is derived from the principle formulated by the Italian economist Vilfredo Pareto (1848 to 1923), who suggested that, in general, 20% of the items in a group can account for 80% of the activity of the group. For example, it may be found that 20% of consumers account for 80% of sales. In practice the 20:80 proportion is not fixed, but these figures conveniently illustrate the principle. In particular cases, the precise proportions would have to be calculated from field data.

Clearly, if it transpired that 20% of a group of buildings used 80% of the total energy of the group, knowledge of this would be essential in setting energy management priorities. The analysis is performed by ranking the buildings in order of decreasing energy consumption. The proportion of the total energy consumption due to the highest user is then evaluated. The procedure is repeated for the next highest and so on. The resulting table can be used to derive the coordinates of a Pareto curve which shows how many buildings account for any given percentage of the total energy consumption of the group. If energy data is not available cost data can be used to derive the curve, although this may result in an altered distribution if the effects of volume discounts or the unit cost of different fuels are significant.

Pareto curves for the energy use of four different groups of buildings are shown in Figure 1. The Pareto principle is illustrated by the buildings of the CBD of Wellington, New Zealand. In this case, 85% of the energy is consumed by 20% of the buildings. It may also be seen that the dominance of large users is less for the other groups. In the case of the Auckland Region, 20% of the commercial buildings use 77% of the energy. This difference may, along with other factors, reflect the different sampling methods used in studying the two groups. The Wellington CBD study included all the buildings in central Wellington while the Auckland study[21] used a 5.4% sample of buildings in a much larger area. Hence, the dominance of ''large'' CBD buildings is not so pronounced in the Auckland study.

In contrast to the two groups of commercial buildings, Curves 3 and 4 in Figure 1 show a lower fraction of total energy costs and consumption resulting from the schools with high energy bills. The top 10% of Wellington primary schools use only 29% of the total energy.[22] This probably reflects the more ''homogeneous'' nature of a group of schools. The ranges of size and energy consumption, and therefore energy costs, are generally lower than for groups of commercial buildings.

Pareto curves provide a useful focus for an energy management program. The greatest reduction in energy consumption within a group of buildings will generally be achieved by concentrating on the larger consumers. In addition, these are the buildings which are most likely to yield cost-effective energy savings, since they will tend to yield greater energy and dollar savings for a given expenditure on controls, maintenance, etc.

The Pareto curve allows assessment of the effect of energy savings by the large consumers on the consumption by the group as a whole. For instance, Figure 2, which is derived from the Pareto curves of Figure 1, indicates the average reductions in energy consumption among the larger consumers required to yield an overall 10% saving in each of the four groups of buildings.

In the Wellington CBD a 14% reduction in energy consumption by the top 10% of consumers (108 buildings) would be required to reduce the total energy consumption of all 1057 commercial buildings by 10%. Alternately, the same result could be obtained by an average reduction of 25% in the 26 buildings with the largest energy consumption.

By contrast, the same strategy applied to the group of schools would be less effective. A 27% reduction in energy consumption among the top 10% of schools would be required to reduce energy consumption in the whole group by 10%. If a maximum average reduction of 14% is considered feasible, then this must be achieved in 28% of the larger consuming schools to effect a 10% reduction in the whole group.

The curves of Figure 2 may be used for cost-benefit analyses of different energy conservation measures.[23] The costs of various conservation schemes may be plotted on the same axes, enabling the evaluation of the energy savings resulting from a scheme of a given cost. Hence it would be possible to determine the optimum allocation of resources within a given energy conservation program.

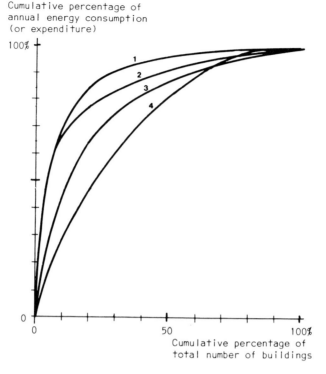

Cumulative percentage of
annual energy consumption
(or expenditure)

FIGURE 1.   Pareto curves of annual energy consumption (or expenditure) for four groups of buildings. (1) Wellington CBD (1057 commercial buildings);[3] (2) Auckland Region (707 commercial buildings);[21] (3) Gloucester County Council (330 schools);[23] (4) Wellington Region (233 primary schools).[22] (Reproduced by permission of the Institution of Professional Engineers of New Zealand).

B. An Index of Energy Conservation Potential

The Index of Energy Conservation Potential (IECP) is a numerical indicator which allows a group of buildings to be ranked. The purpose of the ranking is to identify, in order, those buildings which may be expected to yield the largest energy savings.

The Area Energy Use Index (AEUI), described in detail in Chapter 2, is itself a simple measure of the energy efficiency of a building. However, a measure of energy efficiency does not in itself indicate the potential cost effectiveness of energy-reduction measures. By comparison with large consumers, small ones tend to have lower potential for big reductions in energy use and cost, simply because less energy is involved. In addition, the costs of many energy conservation opportunities are largely independent of the actual consumption. For example, the installation of controls on a high energy flow usually costs less per unit of energy involved than for a small flow. Hence, energy conservation opportunities are usually most cost effective when applied to large users of energy.

To account for the monetary benefits of applying energy conservation measures to these consumers, the energy use index should be combined with the annual cost of energy for the building.

Bearing these factors in mind, the following Index of Energy Conservation Potential (IECP) is proposed:

$$IECP = (Energy\ Use\ Index) \times (Annual\ Total\ Energy\ Cost)$$

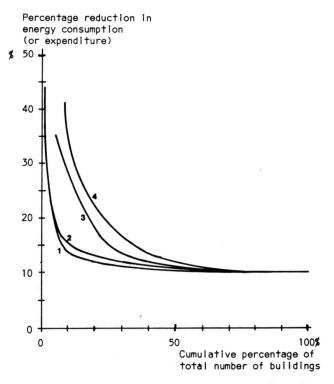

FIGURE 2. Reduction in energy consumption (or expenditure) required to achieve an overall saving of 10% for the four groups of buildings of Figure 1. (1) Wellington CBD (1057 commercial buildings);[3] (2) Auckland Region (707 commercial buildings);[21] (3) Gloucester County Council (330 schools);[23] (4) Wellington Region (233 primary schools);[22] (Reproduced by permission of the Institution of Professional Engineers of New Zealand.)

In the absence of further information, efficiency (in the form of the Energy Use Index) and cost have been given equal weighting.

The Energy Use Index may be refined to take account of the annual "effectiveness" of the energy use of each fuel. The "effective" energy of a fuel is that part of the energy delivered to a building which actually provides a useful service, once conversion losses have occurred. For example, the following average efficiencies could be assumed for the principal fuels used in buildings: electricity, 1.0; gas, 0.8; oil, 0.7; and coal, 0.6. In practice these efficiencies will vary widely between different heating systems and for the same heating system operating at different loads. However, further refinement of the weighting factors could only be achieved by detailed measurement of individual systems.

If the Energy Use Index for an office building (say), is based on effective energy and at the same time normalized by occupant-hours, rather than the more general AEUI, then the Index of Energy Conservation Potential is given by:

$$\text{IECP} = \frac{E + (0.8)G + (0.7)O + (0.6)C}{\text{Sum of the hours per year worked by the occupants}} \times \text{Annual Total Energy Cost}$$

where E, G, O, and C are the annual consumptions (in consistent units) of electricity, gas, oil, and coal, respectively.

The objective is to lower the value of IECP, either by reducing annual energy requirements for a given level of occupancy, or by increasing occupancy for a given level of energy use. The use of the normalizing factor "occupant-hours" avoids separating the energy management function from other aspects of building management such as building operating hours and space allocation.

Until further validation of the IECP takes place, it has been assumed that Energy Use Index and Annual Total Energy Cost have the same weighting. A building which has either double the energy cost or double the energy use of another building, will have double the value of IECP.

1. Applying the Index of Energy Conservation Potential

The IECP can easily be weighted to take account of other building characteristics such as HVAC type, age, etc., if experience shows these to be correlated with cost-effective energy conservation. For instance, central heating systems are likely to provide more worthwhile conservation opportunities than smaller unit systems, because of the larger energy flows involved at the central plant. With experience it may become possible to quantify these characteristics and then use them as further weighting factors.

The index is not normalized for climatic effects and is only valid for comparison of buildings in the same locality, over the same time period. Normalization of total annual energy requirements, to account for climatic variations, is not possible without first separating out the nonclimate related fraction of the energy requirements of a building. However, the use of the index within a given region would enable meaningful comparisons of buildings to be made over the same time period. The data on energy consumption and costs would normally be taken from suppliers invoices.

The result of applying the IECP to a group of 10 buildings, using AEUI as the Energy Use Index, is shown in Table 1.[25] The IECP provides a different ranking order from that obtained using either annual total energy cost or AEUI. For example, the building ranked 6th by IECP value is ranked 25th by annual total energy cost and 4th by AEUI. The IECP can also rank a building higher than either annual total energy cost or AEUI (e.g., buildings 1 and 10 on the IECP ranking). The IECPs for a group of buildings provide an indication of those buildings which offer the greatest potential for cost-effective energy savings. The buildings with the largest IECPs should be examined first. Clearly, the IECP should not be used indiscriminately; the information it provides is only a first step. However, we believe that its use will make it easier to take that first step, and at the same time ensure that the step is in the right direction.

V. MANAGEMENT TECHNIQUES FOR INDIVIDUAL BUILDINGS

In general, when a building has been selected for a detailed energy audit, the only indicator of past consumption will be the billing records of the energy suppliers. These records are useful, not only for calculating Pareto curves and energy performance indexes to initiate management action, but also for providing a base line to *monitor* energy consumption over consecutive periods of time. By providing a historic view of building consumption, past energy records complement a ''walk through'' audit of the building. In addition, analysis of past records can indicate anomalies in the behavior of energy-consuming systems and guide the energy manager to a more detailed investigation of such anomalies.

Two basic techniques for identifying the energy consumption characteristics within individual buildings are described. The first technique — producing energy and power profiles — characterizes energy use over time, and aids estimation of individual system loads, detailed performance profiles, and peak demands. The second technique — performance line analysis — can be used to establish the magnitude of temperature-dependent loads within a building and to detect major changes in the performance of the thermal environmental control systems.

Both techniques can be applied at more than one level, namely, building group level, individual buildings, or to systems and components. They provide an orderly and logical

Table 1
RANKING ORDER FOR THE "TOP TEN" OF A GROUP OF 100 GOVERNMENT BUILDINGS IN WELLINGTON, N.Z.: COMPARISON OF IECP, ANNUAL TOTAL ENERGY COST, AND AEUI

IECP Value		Ranking order		
10^6 MJ m^{-2}$	10^6 Btu ft^{-2}$	IECP	Annual total energy cost	AEUI
38.3	3370	1	3	3
29.6	2605	2	9	1
21.0	1850	3	1	19
16.7	1470	4	2	16
10.8	950	5	20	2
6.1	540	6	25	4
5.4	475	7	6	13
3.9	343	8	14	7
2.5	220	9	38	5
2.0	176	10	12	21

Reproduced by permission of World Energy Engineering Congress sponsored by the Association of Energy Engineers, Atlanta, Ga.

approach to the identification of possible conservation opportunities. The actual technical procedures for implementing energy conservation opportunities have been well documented by others[1, 26] and will not be detailed here.

A. Load Profiles

Load profiles are plots of energy or power vs. time. They normally range from instantaneous demand measurements, made on chart recorders connected to flow meters, watt meters, etc., to graphs of annual consumption drawn from energy meter readings or monthly suppliers' invoices. Graphs of this type, with daily to annual time increments, are the usual method of monitoring energy consumption. However, they are commonly and unjustifiably used as measures of performance. Causes are often assigned to variations in consumption during each period, without knowledge of the magnitude of each end use within the profile, and without any attempt being made to normalize the profile with respect to the many other factors influencing consumption (the use of normalizing variables to measure energy performance will be described in the next section).

Despite the limitations of load profiles as stand-alone measures of performance, when used in conjunction with field surveys of equipment operation, they can help to quantify the end uses of energy. Measurements of power for each item of plant can be combined to produce an estimated total power profile, which can then be compared with the actual power profile. Usually the procedure is iterative; an initial comparison is made and then measurements or estimates are refined until a close correspondence is achieved. Figure 3 is an example of an end-use load profile produced by this technique. To aid energy management decisions, energy use is subdivided into the functions it supports. For instance, lighting consumption is separated into library, lecture theater, office, and cleaning requirements in this example.

Load profiles also indicate the validity of extrapolating spot measurements to estimate consumption over longer periods.[25] Take, for instance, the hourly load profile of electricity use in an office, over a 4-day period, shown in Figure 4. First inspection suggests that because the profile has a repetitive daytime pattern, the end uses which comprise it (lighting,

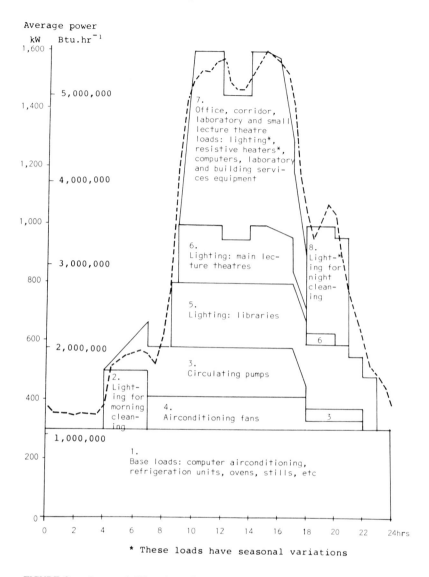

Average power

FIGURE 3.   Average half-hourly profile at a university (weekdays during winter). (Reproduced by permission of the Institution of Professional Engineers of New Zealand.)

office equipment, supplementary heating and cooling) might also have a fixed daytime pattern throughout the year. However, the profile for the same office, based on bimonthly energy consumption readings, shows a major seasonal variation (Figure 5). From this it can be concluded that, in this case, an energy audit establishing the energy consumption of different end uses within the office over a 1-week period is not sufficient for estimating the annual consumption due to each end use.

Because of the different periods between readings it is more accurate to normalize consumption by the number of working days per period to obtain average energy consumption per working day than simply to plot energy consumption between readings. The resulting data can be used to detect seasonal variations. Three years or more of past records should be used to establish whether seasonal trends exist.

The causes of seasonal variation must be interpreted with caution. In the case of electricity, for example, these may be related to one or more of the following:

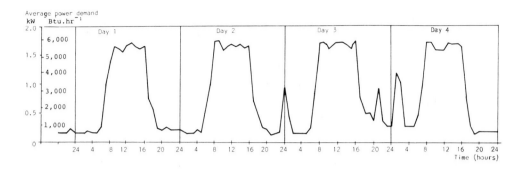

FIGURE 4.   Hourly load profile of electricity use in an office (measured over a 4-day period). (Reproduced by permission of the Institution of Professional Engineers of New Zealand.)

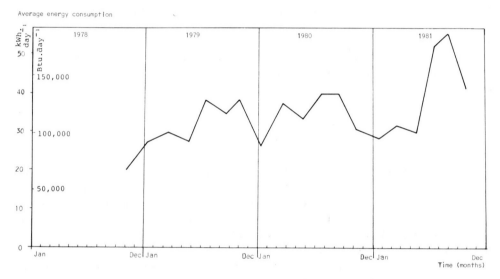

FIGURE 5.   Bimonthly profile for the same electricity consumer as in Figure 4 (from suppliers records over 3 years). (Reproduced by permission of the Institution of Professional Engineers of New Zealand.)

- Changes in working hours throughout the year (both times and duration)
- Changes in operating periods for heating and cooling plant
- Temperature related operation of heating and cooling plant
- Availability of daylight

Generally the causes of seasonal variation can only be determined by establishing the changes in each end use. This requires a detailed survey of energy-consuming services and their controls.

Load profiles can illustrate how energy consumption varies between monitoring intervals. If, for example, hourly readings are taken, these can be used to determine the base load component of the total daily energy consumption. By contrast, monthly readings (say from supplier's records) can be used to check the validity of extrapolating hourly or daily readings to provide estimates of energy consumption over longer periods. In addition, once estimates of the demands and operating times of individual loads have been made, the load profile for each fuel provides a useful, although not necessarily sufficient, check of these estimates.

B. Performance Line Analysis

To monitor the operating performance of the energy-consuming systems of a building, it is necessary to take account of uncontrollable influences which affect energy requirements, so that consumption can be compared over two or more time periods.

Once this has been done, the "corrected" or "normalized" energy use can then be compared with earlier periods. This technique can be used to measure the energy performance of single buildings over a period of time. It is useful both as an evaluative tool to assess the effects of plant modifications and operational changes and as a diagnostic tool to detect plant malfunctions and the need for remedial maintenance.

The major application of this type of monitoring is to provide feedback on the effectiveness of an energy management program and to direct maintenance personnel to apparent anomalies in building energy use. Unlike the energy indexes described earlier, it is not intended as an interbuilding performance measure.

The essential elements or requirements of a performance monitoring technique are

- The identification of uncontrollable variables which significantly affect energy consumption
- Methods of normalizing energy consumption with respect to these variables
- Determination of the frequency with which to monitor energy consumption and the variables which affect it (e.g., daily, weekly, monthly)
- Methods for detecting significant changes in energy consumption

The following sections describe a simple monitoring technique which may be used for measuring the performance of space heating systems.

1. Normalizing Space-Heating Energy Consumption

Normalizing factors adjust energy consumption for the uncontrollable influences, such as outdoor temperature, sunlight, wind, etc., which affect it. To a first-order approximation, the heating requirement of a building is directly proportional to the difference between inside and outside temperature. Solar radiation and wind speed also affect energy consumption, though their effects are often secondary.

Because of the dominant effect of outdoor temperature, both average temperatures and degree days have been used for normalizing heating energy requirements. This permits consumption over different time periods and under different climatic conditions to be compared.

The average of the maximum and minimum daily temperature has been used by the U.K. PSA in monitoring the performance of building heating systems.[27] Average temperatures have also been used in Part 4 of the CIBS Building Energy Code[28] and in the energy management program for state buildings in Iowa.[29]

Degree days have been widely used as a simple method of predicting the space-heating energy requirements of a building.[30] They have also been used, in both conventional[31] and modified[32] forms, as normalizing variables.

If average temperatures are used for normalizing space-heating energy consumption, then a straight line of negative slope should result from plotting energy vs. average temperatures (per period). The relationship is represented graphically in Figure 6, which shows three different performance lines representing different levels of "performance". Although the effects of solar heat gain, wind, thermal mass, plant efficiencies etc., are ignored, Livesey and Taylor[33] report that a good approximation to a straight line has been obtained for well-maintained heating systems under effective control. The "ideal" performance line cuts the baseload below the point where the average outside air temperature equals the design internal temperature. Variations from the "ideal" line, or excessive scatter, indicate poor performance.

The line itself is fitted to a set of readings by the statistical technique of regression analysis. Performance lines have been used to monitor all the large buildings (i.e., those with boiler capacities exceeding 150 kW or 500,000 Btu hr^{-1}) for which the PSA is re-

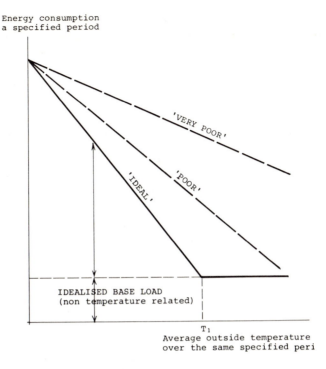

Energy consumption
a specified period

'VERY POOR'

'IDEAL'

'POOR'

IDEALISED BASE LOAD
(non temperature related)

T_1
Average outside temperature
over the same specified peri

FIGURE 6.   Idealized performance lines for building heating systems. (Note that T_1 is the threshold average temperature below which heating energy is used. For the ''ideal'' line, this might be expected to be less than the inside design temperature).

sponsible. Its highly successful monitoring program was based on weekly readings of heating fuel consumption and the weekly average of daily maximum and minimum temperatures.

The degree day is another temperature-based normalizing factor. It is commonly defined as the difference between the average daily temperature and a preselected base temperature. The daily values are added to give the degree-day total for a given period. Space-heating energy consumption is then plotted against the degree-day totals. In this case the expected relationship is a straight line of positive slope cutting the energy consumption axis at the baseload value.

Both these types of monitoring analysis were carried out for eight multistory buildings with a wide range of HVAC systems in Wellington, New Zealand, using energy consumption readings covering annual periods from 1976 to 1979.[25] Most buildings had monthly gas readings from a supply authority, in some cases weekly readings had been taken by the consumers, while a few utilized bimonthly electricity readings from the supply authority.

Strictly speaking, the following comparison of average temperature and degree days is only applicable to mild climates similar to that of Wellington, which averages 2074°C-day (3733°F-day) with an 18°C (65°F) base temperature, 2020 hr of sunshine well distributed throughout the year, and daytime design temperatures of 23.3°C (74°F) in summer and 3.3°C (38°F) in winter at the 1% level.

The coefficients of determination (r^2) of the ''best-fit'' performance lines, using both average temperatures and degree days, are listed in Table 2. A more detailed explanation will be given in a later section, but for the moment it may be assumed that, in general terms, an r^2 value can be interpreted as the percentage association between two factors.

Overall, it was found that average temperature was more highly associated with space-heating energy consumption than degree days. However, in regions with higher summertime

Table 2

PERFORMANCE LINE COEFFICIENTS OF DETERMINATION (r^2) FOR THE HEATING SYSTEMS OF EIGHT BUILDINGS

				Coefficients of Determination(r^2)			
					Degree-day base temp(°C)		
Building	Fuel type	Year	Data points	Average temp	18	19	20
A	Gas	1976	10	0.882	0.882	—	—
		1977	7	0.816	0.826	0.826	0.826
		1978	9	0.921	—	—	—
		1979	9	0.880	—	—	—
B	Gas	1976	11	0.862	0.857	0.862	0.862
		1977	11	0.970	—	0.970	0.969
		1978	11	0.970	0.881	0.962	—
		1979		0.924	—	0.910	—
C	Gas	1977	11	0.727	0.654	0.717	0.717
		1978	11	0.570	0.202	0.400	—
		1979	11	0.887	0.941	0.919	0.915
D (monthly)	Gas	1979	11	0.933	0.927	0.937	—
D (weekly)	Gas	1979	50	0.848	0.668	0.813	0.824
E	Gas	1977	11	0.976	0.968	0.976	0.976
		1978	11	0.936	0.645	0.890	0.930
		1979	12	0.158	—	0.126	—
F	Gas	1977	11	0.714	0.674	0.707	—
		1978	11	0.734	—	—	—
		1979	11	0.932	—	—	—
G	Gas	1977	10	0.528	—	—	—
		1978	10	0.642	—	—	—
		1979	13	0.628	—	—	—
H	Electric	1976	50	0.692	—	—	—
		1977	50	0.803	—	—	—
		1978	50	0.726	—	0.029	0.307

temperatures and more variable between-season temperatures than Wellington, a degree-day measure may be preferable, since it avoids the possibility of averaging a low temperature when heating is required, with a high temperature when cooling is required, in the same interval. This effect can result in two periods, with the same average temperature, having very different heating requirements.

The modification of the degree-day method described by Mayer and Benjamini[32] takes into consideration the distribution of temperatures over a typical day. Temperatures are assumed to vary linearly with time between the daily maximum and minimum. In addition, these modified degree days replace the conventional degree-day base by a reference temperature which can be varied to fit the data. It allows, in crude form, for the possibility of different interior temperatures, solar heat gains, heat sources, and heat losses affecting the thermal balance of individual buildings. However, the method requires both the minimum and maximum temperature for each day.

The discussion in the following sections is based on analysis of energy consumption using average temperature as the normalizing variable.

2. Monthly Gas Consumption Readings

The performance lines generated for buildings with monthly gas readings indicated a high association between heating energy use and average temperature. The corresponding r^2 values are listed in Table 2. As an example, in the case of Building E in 1977, 97.6% of the variation in heating energy use can be associated with the average temperature ($r^2 = 0.976$).

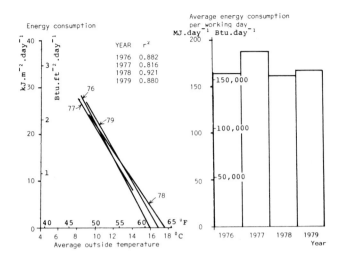

FIGURE 7. Performance lines and annual energy consumption figures for Building A heating system. (Reproduced by permission of the Institution of Professional Engineers of New Zealand.)

The high associations which occur for each year (but not necessarily for the same building) indicate that average temperature can account for a large portion of the variation in space-heating energy consumption. The high associations and similar performance lines which occur for the same buildings over different years such as Building A, (Figure 7) suggest that average temperature is capable of allowing for climatic variations and indicating consistent performance.

Several of the buildings listed in Table 2 have relatively low percentage associations and/ or significantly different performance line slopes between years (for example Building C 1978 and Building B 1976). Field studies have shown that these variations are associated with major changes in building operations. For example, in the case of Building C (Figure 8) the ownership and some of the tenants changed between 1977 and 1979 resulting in the building being operated for fewer hours at night. This produced a much lower annual energy consumption and a corresponding drop in the energy intercept of the performance line after 1977. The low association in 1978 ($r^2 = 0.57$) indicates the changing operation of the building. As 1977 and 1978 were characterized by low r^2 values and the confidence intervals on both slopes are wide, the only conclusion that can be drawn about these two lines is that they are both showing a very poor response to outside temperature. This is characterized by their very high intercepts with the temperature axis (the system has no baseload in this case).

By contrast, the 1979 performance line for Building C shows a better response to temperature, a higher r^2 value, and a much reduced consumption over the whole temperature range. The line is different from the other two at greater than the 1 in 20 (or 95%) level of significance. Building C had separate air handling units with heating and cooling coils serving different zones. During two surveys of the plant room, conducted at semiannual intervals, some of the air handling units had heating and cooling coils operating simultaneously. If this condition had occurred for a number of years it would have caused the large scatter observed.

Other buildings gave a set of data points which describe either a convex curve or two lines of different slope (see Figure 9). This phenomenon has only been observed with gas consumption and could indicate either poor part-load boiler efficiency or excessive on-off cycling. However, for the monthly readings this curvature was not statistically significant at the 1 in 20 level.

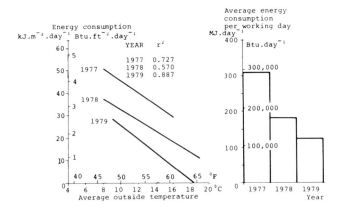

FIGURE 8.   Performance lines and annual energy consumption figures for Building C heating system.

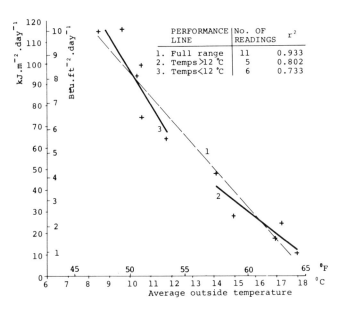

FIGURE 9.   Performance line analysis of Building D heating system using monthly readings (1979).

3. Weekly Gas Consumption Readings

   The 1979 weekly gas readings for the Building D heating system were analyzed (see Figure 10) and the results compared with the performance line data for the same system using monthly records (see Figure 9). It will be seen that the r^2 value for the 50 weekly readings is 0.848, which is less than the r^2 value of 0.933 for the 11 monthly readings. The lower r^2 value for the weekly readings does not necessarily imply that the linear relationship between energy consumption and temperature is any less significant. The significance of the r^2 value is dependent on the number of data points from which it is derived, as will be explained in detail in a later section.

   In Figure 9 the gas consumption varies by a maximum of 18 kJ m^{-2} day^{-1} (1.6 Btu ft^{-2} day^{-1}) about the best-fit line for monthly readings. The corresponding maximum variation for weekly readings (Figure 10) is 41 kJ m^{-2}day^{-1} (3.6 Btu ft.$^{-2}$ day^{-1}). Although the

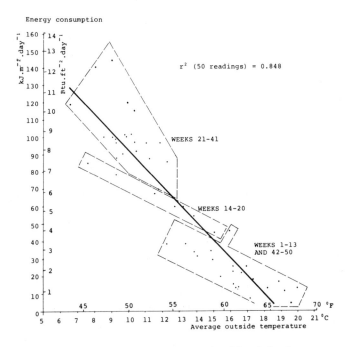

FIGURE 10.   Performance line analysis of Building D heating system using weekly readings (1979).

performance line, based on monthly readings, has a high correlation coefficient and relatively low scatter, the plot of weekly readings shows very large weekly variations are occurring, which have been disguised by the monthly aggregation of data. Conversely, the monthly data best illustrates basic trends.

In Figure 9, differences occur in the slope of the lines through the monthly data points, above and below 12°C (54°F). However, these differences are not significant at the 1 in 20 level. Nevertheless, when the same analysis is repeated for the weekly data of Figure 10, the two lines are significantly different at the 1 in 20 level. The weekly readings indicate a significant difference between the operation of the heating system, for average temperatures either side of 12°C (54°F).

In the weekly plot the readings have been separated into consecutive periods. Weeks 1 to 13 and 42 to 50 cover high-temperature operation and show a moderate amount of scatter. Weeks 14 to 20 indicate a strong linear relationship with temperature. Weeks 21 to 41 show very litte temperature-related operation. These variations are systematic and imply that even heating systems with good performance lines may have controls that are sometimes inefficient. In this case the building was heated by two boilers, one of which was set to operate whenever heating was required (i.e., below an outdoor threshold temperature) while the other was kept off, or set to operate at lower outdoor temperatures when the building demand was expected to be greater than the capacity of the first boiler. The weekly data reveals all these modes, whereas the monthly readings aggregate consumption occurring in different modes and hide these operational differences.

4. Bimonthly Electricity Consumption Readings

Most of the performance lines based on electricity consumption data consisted of only five or six points per year, as a result of a bimonthly meter reading period. Thus the range of average temperatures per reading period is generally smaller than for the monthly gas readings. These factors reduce the potential of performance line analysis to separate temperature-independent consumption from temperature-dependent consumption and to reliably indicate the type of variation in the latter with average temperature. This problem is compounded by metering and wiring arrangements which typically include lighting, office equipment, and heating on the same circuit. The sum of the lighting and equipment loads is often as great as the heating load and is generally not constant. Consequently, the variations in these loads reduce the reliability of any observed variation in total energy consumption with outside temperature.

The performance lines generated from electricity readings indicated these effects. Despite the fewer data points available for each year, the percentage association was generally lower than for the lines based on monthly gas consumption. This indicated the need to quantify the end uses to establish whether performance line analysis is appropriate. Where the magnitude or the extent of variations in baseload is large, the resulting degree of scatter about the performance line provides a poor basis for indicating system performance.

5. Statistical Analysis of Performance Lines

So far, the characteristics of performance lines have been described in mainly qualitative terms. This section outlines the statistical tests of annual performance lines which can be used to determine the level of significance of coefficients of determination (r^2 values) and differences between slopes and intercepts. These tests are essential for establishing whether apparent differences in performance lines are statistically significant.

The correlation coefficient (r) indicates the goodness of fit of a line fitted by the method of least squares. It is the most widely used measure of the strength of the linear relationship between two variables (say X and Y).

The coefficient of determination (r^2) provides a measure of the degree of linearity (or association) between X and Y. By itself, the coefficient of determination tells us little about the significance of a straight line through a set of data points. It must be interpreted in relation to the number of observations on which it is based. A higher coefficient of determination is required for a small number of observations than for a large number, if the odds of a straight line relationship not being due to chance are to be similar. Table 3 contains values of r^2 which must be exceeded for a given number of observations (n) before it can be concluded that a linear relationship is significant at a predetermined level. For instance, if the performance line generated from six data points (e.g., six bimonthly fuel measurements) has an r^2 value 0.700, then the odds are better than 20 to 1 that the relationship exists. However, it would not be possible to say that the linear relationship exists at the 1 in 50 level of significance; the value of r^2 would have to be 0.778 or more (see Table 3) for this to be the case. For an annual performance line based on weekly readings (52 points) r^2 need only equal 0.075 for the odds to be 20 to 1 against the correlation being due to chance.

The percentage variation of the energy consumption measurements which can be attributed to differences in temperature (as opposed to other factors or to chance variation) is given by 100 r^2. However, the coefficient of determination gives no indication of the actual relationship between energy consumption and temperature; it simply indicates the likelihood that a linear relationship exists and gives an estimate of its strength.

The actual r^2 value which indicates good heating system performance for a particular building cannot be prescribed without detailed knowledge of the type of controls within the building and how consistently it is operated (e.g., whether the weekly operating hours are constant). A fluctuating baseload is likely to introduce "noise" which will lower the coefficient of determination regardless of how consistently the space-heating system is operated. The energy consumption of completely automated space-heating systems is likely to show better correlation with outdoor temperature than would manually operated systems.

6. Applying the Performance Line Technique

The performance line technique can be used in two main ways. First, as a continuous

Table 3
CRITICAL VALUES OF THE COEFFICIENT OF DETERMINATION (r^2) AT THREE LEVELS OF SIGNIFICANCE

No. of data points (n)	Critical values of r^2 for the given levels of significance		
	1 in 20	1 in 50	1 in 100
5	0.771	0.872	0.920
6	0.658	0.778	0.841
7	0.569	0.694	0.766
8	0.500	0.623	0.696
9	0.444	0.563	0.637
10	0.399	0.511	0.585
11	0.362	0.469	0.540
12	0.332	0.433	0.501
17	0.232	0.311	0.367
22	0.179	0.242	0.288
27	0.145	0.198	0.237
32	0.122	0.167	0.202
37	0.106	0.145	0.175
42	0.092	0.128	0.154
47	0.083	0.114	0.138
52	0.075	0.104	0.125

Note: For a given number of data points (n) the coefficient of determination (r^2) must exceed the values shown if the odds of a straight line relationship being due to chance are to be less than 1 in 20, 1 in 50, or 1 in 100.

"on site" check, with periodic readings being plotted onto an existing performance line, to establish whether consumption differs from that predicted by the performance line. Used like this, the performance line simply indicates a change in performance, but not its cause. The second use involves comparing performance lines for two different periods to observe any systematic change in the temperature related operation of the system.

The two uses are complementary. The first technique is suitable as a manual check for plant operators, enabling them to spot any change in operation and investigate its causes. The second technique is more suitable for computer analysis. It can identify systematic changes in performance and diagnose possible causes, albeit over much longer time periods. The aim of this type of analysis is to keep a check on an established system in operation, and to ensure that it continues to operate satisfactorily.

Application of the performance line concept may not be useful when a metered fuel has two or more end uses, each of which varies with respect to a different independent variable. This applies, for example, with electricity where resistive heating and lighting loads are often metered together. The lighting loads may be highly dependent on hours of daylight, cloud cover, and so on, but not particularly related to the average outside temperature. A similar situation can occur in institutions such as hospitals where steam or hot water is used for both space heating and cooking. The latter requirement varies according to the menu. In these situations the performance line should not be used unless the effects of these "nonheating" end uses are known.

a. Data Requirements

Performance lines may be calculated using a minimum of five data points, but it is preferable to have at least ten. Weekly readings provide information covering greater ranges of both energy consumption and temperature variation than do monthly readings, and the corresponding performance line indicates the performance of the system over a wider range of outdoor temperatures.

The elapsed times between readings, used for calculating a performance line, should be of similar duration. Performance line comparisons should only be made if they are based on readings taken over similar time intervals. Bearing in mind the above considerations, weekly readings are best used for generating, say semiannual performance lines (e.g., for winter or summer) and monthly readings are best used for generating annual performance lines.

Generating a base-year performance line from monthly readings (e.g., made by a supply authority) and using this to assess the weekly performance of a heating plant is unlikely to be valid. It would require extrapolation outside the range of actual readings. Systematic differences exist between the slopes and intercepts of performance lines based on average temperature, for different frequencies of observation.

b. Confidence Intervals

It must be borne in mind that the coefficient of determination and its significance indicate only the degree of confidence with which it can be asserted that a linear relation exists between energy consumption and average temperature. It does not indicate how accurately the performance line equation estimates the actual relationship between energy consumption and temperature, or whether in fact the relationship is predominantly linear.

However, the performance line can be used to compare energy use over different periods and to determine whether systematic variations have occurred. To do this, the slope and intercept of one performance line are compared with those of another, using an appropriate statistical test. For a certain level of significance the confidence intervals can be computed for each slope and intercept.

The confidence intervals for the slope (β) and energy intercept (α) of the actual equation, for which the performance line is a sample estimate, are

$$\text{Actual slope } \beta = b \pm \frac{t_{(1 \text{ in } c)} \, Se}{[\{n(\Sigma x^2) - (\Sigma x)^2\}/n]^{0.5}}$$

$$\text{Actual intercept } \alpha = a \pm t_{(1 \text{ in } c)} \, Se \left[\frac{1}{n} + \frac{n(\bar{x})^2}{n(\Sigma x^2) - (\Sigma x)^2} \right]^{0.5}$$

where n = the number of data points, a = the estimated energy intercept (based on measurements), b = the estimated slope of the performance line (based on measurements), $t_{(1 \text{ in } c)}$ = the t-statistic for the level of significance chosen, 1 in c = the level of significance (e.g., 1 in 20, 1 in 50, 1 in 100, etc.), Se = the standard error, and $\bar{x}$ = the mean value of x.

Rather than compare the slopes and intercepts independently, it is sometimes desirable to determine the confidence bands within which a second performance line may fall without being considered significantly different. To do this, the confidence limits of the energy consumption estimate are calculated for any given temperature and plotted onto the performance line diagram. The resulting band defines the area in which other performance lines will lie, if they are not significantly different. The confidence limit for $\alpha + \beta x_o$ (i.e., the true energy consumption at temperature x_o is

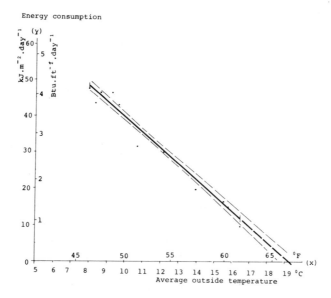

FIGURE 11. Performance line and confidence band for Building B heating system (1977). Line of least squares fit, y = 83 − 4.36 x (SI Units); coefficient of determination, r^2 = 0.970; standard error, Se = 2.63 (SI Units).

$$(a + bx_o) \pm t_{(1 \text{ in } c)} \, Se \left[\frac{1}{n} + \frac{n(x_o - \overline{x})^2}{n(\Sigma x^2) - (\Sigma x)^2} \right]^{0.5}$$

The confidence band about the performance line is always narrowest at the mean temperature and widest at extreme values. Figure 11 shows the points for the heating system of Building A during 1977, together with the best-fit performance line and its confidence band (i.e., the line through the confidence limits of the energy consumption estimate).

VI. A STRATEGIC FRAMEWORK FOR BUILDING ENERGY MANAGEMENT

As will be readily appreciated, buildings that were not designed in an energy-conscious manner are unlikely to have sufficient provision for detailed monitoring of energy performance. What we have described in the last few pages are diagnostic methods, which make use of the limited energy data available from supplier records, or obtainable by using portable instruments.

To complete the picture, we present a strategic framework for building energy management in which diagnostic methods are matched to the appropriate level of decision making. There are several levels at which decisions can be made regarding the management of energy use in buildings. These may be broadly classified as follows:

- The national level
- The building group level
- The individual building level
- Specific energy-consuming systems and components

The decisions required at each of these levels will now be considered.

A. The National Level

Clearly, government policies have a major impact upon all the other levels. These policies will tend to be directed towards operating within the constraints described at the outset of this chapter. Ideally, the consumption of each fuel for a particular end use is required (e.g., oil and gas for space heating, electricity for chilling, ventilation, lighting, etc.). Once this is known, the development and evaluation of strategies for energy demand management can commence. At this level, energy management involves matching fuels with their most appropriate end uses. This will vary from region to region according to such characteristics as the local resource base, fuel availabilities and fuel production, and distribution costs.

Besides such macro-energy planning, government policies can directly influence energy use through the provision of incentives which are available to building owners. These include grants, low-interest loans, tax concessions, and so on for energy conservation measures such as the installation of more efficient equipment or for undertaking an energy audit to identify energy conservation opportunities. Regulatory controls may also be applied (especially during periods of energy shortages) to achieve more immediate energy reductions. These include specified operating policies for heating and cooling or energy consumption controls. Before implementation, any government policy should be considered in relation to the constraints and opportunities that exist, including those discussed in the opening sections.

To predict the impact of such measures, information is required on the breakdown of consumption between subsectors (e.g., hospitals, offices, schools, houses) and then the distribution of consumption within each subsector. This can be used to rank consumers of different fuels. Such information helps to quantify the likely impact of decisions on fuel price relativities, availability of low interest loans for energy conservation, research development and demonstration priorities, and other such matters.

B. The Building Group Level

For property managers responsible for a large number of buildings, the basic problem is to identify those individual buildings which are likely to yield the most cost-effective savings.

In order to do this, techniques for ranking building energy use, often using existing records, can be used and simple indexes of energy performance prove useful. Pareto analysis and an IECP are most useful at this stage. Generally the performance of building energy systems will leave room for improvement. The first step is to reduce the energy consumption of existing equipment as much as possible by "good housekeeping" before considering more costly and complex methods of energy conservation.

C. The Individual Building Level

In terms of the management of individual buildings, it is most important first of all to quickly detect any deterioration in energy performance and then to rank the various energy conservation opportunities that exist.

In order to monitor performance, it is useful to have a measure of the energy performance of a building which may be simply applied at frequent intervals. Performance line and load profile methods have been described which are applicable at this level. For the ranking of opportunities, it is essential to have a breakdown of the energy end uses in the building; heating, lighting, cooling, transportation, and so on.

D. Specific Energy-Consuming Systems and Components

At this level, the building operation and maintenance personnel must perform the essential roles. However, too much emphasis is often placed on individual components such as boiler plant, chillers and fans, and too little on the performance of the overall system. Again, simple diagnostic tools, capable of rapid and unambiguous interpretation, are required, but these are well documented elsewhere.[1]

Virtually all the methods described have some application at the individual building level. While some methods are most appropriate to national decisions, others apply better at the

	DECISION MAKING LEVEL		
	NATIONAL & BUILDING GROUP	INDIVIDUAL BUILDINGS	SYSTEMS & COMPONENTS
PARETO CURVE	To quantify the significance of major energy consumers		
ENERGY INDICES	To identify energy conservation potential	To allow comparison with similar bldgs To normalize for changes in climate, usage, etc.	
PERFORMANCE LINES	To distinguish temperature dependent consumers	To distinguish temperature dependent loads To detect changes in building performance	To characterize thermal envnt'l control systems and detect changes in performance
LOAD PROFILES	To predict the energy required from suppliers	To detect overall energy use chars. over time and allow estimates of indiv. system loads. To identify periods of peak demand	To investigate detailed performance characteristics

(Left axis label: DIAGNOSTIC METHOD)

FIGURE 12. A strategic framework for building energy management. (Reproduced by permission of the Institution of Professional Engineers of New Zealand.)

systems and components level. Figure 12 summarizes the principal application of each method at each decision making level.

We believe that this strategic framework, together with the methods outlined in this chapter will enable managers at all levels to systematically plan the elimination of energy waste in the most cost-effective manner possible.

REFERENCES

1. **Dubin, F. S., Mindell, H. L., and Bloome, S.,** How to Save Energy and Cut Costs in Existing Industrial and Commercial Buildings, Noyes Data Corporation, New Jersey, 1976.
2. **Hardin, G.,** The tragedy of the commons, *Science,* 1968.
3. **Baird, G., Donn, M. R., and Pool, F.,** Energy Demand in the Wellington Central Business District — Final Report, NZERDC Publ. No. 77, New Zealand Energy Research and Development Committee, Auckland, 1982.
4. 10 Year Buildings Plan for Energy Conservation, Energy Conservation Division, Office of Buildings Management, Public Buildings Service/General Services Administration, Washington, D.C., May 1980.
5. Third Annual Report on Energy Management, Energy Conservation Division, Office of Buildings Management, Public Buildings Service/General Services Administration, Washington, D.C., September 1980.
6. **Lehtinen, M. H.,** New York City's energy conservation program, *ASHRAE J.,* 22, 7, 53, 1980.

7. NYC Agency Energy Conservation Manual, Report UA/DGS/100, Urban Academy for Management Inc, New York, July 31, 1979.
8. **Eisdorfer, H.,** Energy Conservation Guide for School Building Operation, Board of Education City of New York, Division of School Buildings, Long Island City, N.Y., September 1980.
9. **Livesey, P. M.,** National savings, *J. Chartered Inst. Build. Serv.,* 3, 8, 51, 1981.
10. **Livesey, P. M.,** PSA Energy Conservation Group Leader, personal communication, November 9, 1981.
11. **Provan, J. D.,** Energy management, targeting and monitoring in Strathclyde regional council, in Proc. Nat. Energ. Management Conf. 1980, Birmingham, U.K., October 7 to 8, 1980; available from Department of Energy, London.
12. 5th Annual Report of the Energy Conservation Unit, Department of Architecture, Cheshire County Council, Goldsmith House, Hamilton Place, Chester, England, June 1980.
13. Estimated Energy R & D Funding by Provincial Governments in 1976-77, 1977-78 and 1978-79, Report ER 79-5, Office of Energy Research and Development, Energy Mines and Resources Canada, Ottawa, June 1979.
14. Energy Conservation in Ontario Government Buildings — Five Separate Case Study Leaflets, Ontario Ministry of Energy, April 1979.
15. **Armour, A.,** Report on the Proc. Cities Energ. Conf., Toronto, January/February 1980.
16. **Lang, R. and Armour, A.,** Sourcebook — Energy Conservation in Twenty Canadian Cities, Toronto, 1980.
17. Energy Policy for the City — Better Services with Less Energy, Commissioner of Planning and Development, City of Toronto, July 1979.
18. Energy Conservation Progress Report for Period January 1 to December 31, 1978. City of Toronto Board of Education, June 11, 1979.
19. **Baird, G., Brander, W. D. S., and MacFarlane, J. N. W.,** Energy management in New Zealand government buildings, in Proc. 3rd CIB-W67 Symp. Energ. Conserv. Built Environ., Dublin, Ireland, March 1982.
20. Economic and Other Criteria for Energy Conservation in Building Services, Tech. Memo. BS 33, Ministry of Works and Development, Wellington, March 1982.
21. Beca Carter Hollings and Ferner in association with Shaw, R. A., Greater Auckland Commercial Sector Energy Analysis, NZERDC Publ. 45, New Zealand Energy Research and Development Committee, Auckland, May 1979.
22. **Bruhns, H. R.,** Energy Research Group, School of Architecture, Victoria University, Wellington, personal communication, December 1981.
23. **Davis, D.,** Pareto analysis, *J. Chartered Inst. Build. Serv.,* 3, 5, 49, 1981.
24. **Socolow, R. H.,** The Twin-Rivers program on energy conservation in housing-highlights and conclusions, *Energ. Build.,* 1, 2, 1978.
25. **Baird, G. and Brander, W. D. S.,** Energy Conservation in Government Buildings — Report on Stage 2, Contr. Res. Paper 5, School of Architecture, Victoria University of Wellington, February 1982.
26. CIBS Building Energy Code Part 3, Chartered Institution of Building Services, London, 1979.
27. PSA, Monitoring Fuel Consumption of Heating Installations, Technical Instruction (M&E), M79, Department of the Environment, Property Services Agency, Croydon, England, December 1974.
28. CIBS Building Energy Code Part 4, The Chartered Institution of Building Services, London, 1982.
29. **Woods, J. E. and Peterson, P. W.,** Development of a Building Energy Management Program for State Buildings in Iowa, Building Energy Utilization Laboratory, Engineering Research Institute, Iowa State University, Ames, 1978.
30. ASHRAE, ASHRAE Handbook and Product Directory — 1980 Systems, American Society of Heating, Refrigerating and Air Conditioning Engineers, Atlanta, Georgia, 1980.
31. Department of Energy, Degree Days, Fuel Efficiency Booklet 7, Her Majesty's Stationery Office, London, 1977.
32. **Mayer, L. S. and Benjamini, Y.,** Modelling residential demand for natural gas as a function of the coldness of the month, *Energ. Build.,* 1, 301, 1977/78.
33. **Livesey, P. M. and Taylor, B. A.,** Improvement of Energy Efficiency in Existing Office Buildings, Property Services Agency, Croyden, 1980.
34. **Freund, J. E.,** *Modern Elementary Statistics,* 3rd ed., Prentice-Hall, London, 1973.

Chapter 8

BUILDING ENERGY PERFORMANCE — FUTURE CONCERNS

I. INTRODUCTION

This book has examined some of the more important aspects of the energy performance of commercial and institutional buildings. Our main aims have been to explain the consequences of current practices and to provide a basis for the development of new ones. We have attempted to meet these aims by concentrating on the results of studies into the energy performance of existing buildings and by providing a cohesive framework for such studies, now and in the future. We have highlighted the current concerns of nations, owners, designers, and users in relation to building energy performance. In this final chapter, we shall look briefly at the future concerns of each of these groups.

II. MEASUREMENT OF BUILDING ENERGY PERFORMANCE

The development of measures of energy performance, appropriate to the needs of all the people involved with buildings, has been a primary objective throughout this book. Of all the measures that have been proposed, the Area Energy Use Index (AEUI) has gained the most acceptance. It is used in national standards by building owners, by designers, and by energy managers. However, it will be some time before the implications of a given AEUI value will be clear to the users (be they nations, owners, designers, or managers). Time and exposure are needed for people to develop a feel for AEUI values as measures of building energy performance.

In the case of the building designer, for example, a basic understanding of the various measures of energy performance (be they AEUI or any other) is essential. As with so many aspects of building design, the initial decisions tend to have the most far reaching effects. It is at these initial stages that energy performance must be considered, along with all the other aspects of the performance of a building. We hope that the reader will now have some feeling for the various measures of energy performance (the AEUI in particular) and the factors that affect them in the real world. It is now over to readers to test these measures on the buildings for which they are responsible and obtain that deeper understanding that only comes from practical experience. Eventually, experience will provide most of the information needed by those concerned with building energy performance. The practitioner will learn what combinations work and what their energy use consequences will be.

There will always, however, be new projects. The need to design for a new climate, size, activity, energy system, or construction system, for example, will result in the building designer having to seek new information and carry out new analyses.

III. THE EXOGENOUS-ENDOGENOUS CLASSIFICATION

It is in the provision of new information that much of the energy performance research work will be concerned. The interface between design and science contains much that is of interest and which remains largely unexplored. The exogenous-endogenous (exo-endo) classification system outlined in Chapter 2 is an attempt to clarify the problems to be faced at this interface. The classification is an essential first step towards a building design process (in the broadest sense) in which energy performance is well integrated. More work will be needed before the evaluation and analysis efforts of building research are converted into useful information. The designer, for example, needs specific kinds of information at different times in the design process. One task must be to have the right information available at the appropriate time during the design process.

The exo-endo classification system sets up a framework for making the right decisions at the right time. In fact, it prompts questions to be asked at the time in the design process when answers can be most effective. The research information is not yet all available and the design process needs further clarification, but the framework for bringing the two together exists. We see the exo-endo system as providing the means by which energy performance considerations may be integrated into the whole building process, from predesign feasibility studies through to postoccupancy evaluations.

A future concern therefore is for more research into the contexts in which the results of building energy performance studies will be used. It is not enough to produce evaluations of existing buildings or design tools for new ones. Their application must also be carefully studied.

Classification of the information that is available in relation to its point of application in the building process and the person most likely to use it effectively, is a useful first step. Whether or not the classification uses the terminology of the exo-endo system is less important. What is vital is that those people involved in the building process know what questions to ask, and when. It must be possible for them to identify important features of the building, to establish their energy performance repercussions, and to pinpoint the individual decision makers who can influence that performance.

We shall now look at the likely future concerns of those involved in the building process, starting at the national level.

IV. NATIONAL CONCERNS

Studies of energy supply and demand will be of continuing concern at the national level. The results of such studies are needed for energy planning. As mentioned in Chapter 3, detailed assessment of the demand side has been slow due to its diffuse nature. In the case of commercial and institutional buildings, more results of systematic surveys are now becoming available, enabling national and regional models of energy consumption of this sector to be developed and tested. Considerably more work can be expected in this area, leading to more accurate forecasts of the effects of implementing a range of energy conservation policies at the national level.

Energy standards for buildings will continue to be one of the most important manifestations of national concerns. Considerable effort has already been expended, and experience gained, in formulating energy standards for buildings. These attempts are leading to a critical appraisal of the energy implications of existing building standards and a reassessment of the methods used to predict and monitor energy consumption in buildings.

Prescriptive standards have appeared to be the easiest to implement and check, but in practice it is difficult to assess their real impact on energy consumption. Studies are already taking place to check compliance with prescriptive standards. The effectiveness of such standards in reducing energy use must be dealt with in the future.

Performance standards have been formulated by many bodies throughout the world. It has rapidly become apparent that our current knowledge of building energy performance is inadequate. As mentioned earlier, people do not yet have a feel for the various measures of performance involved. In addition, some of the standards — or at least the methods required to check the compliance of a building with them — have become far too complex. They ignore the real needs of the building practitioner, particularly at the initial stages of a project. Several of these standards are already in existence, and it is important that feedback be obtained on their application and their effects in practice, so that they may be developed and improved, or abandoned, as appropriate.

Those formulating the prediction-type performance standard have very quickly come up against the shortcomings of pre-1973 methods of predicting building energy use. The usefulness and limitations of some of these prediction methods are discussed in Chapter 6. New computer-based and manual methods have been developed for the use of designers; it remains to be seen how these methods will fare. In both cases feedback is essential if the methods are to become acceptable design tools. The testing and validation of any new prediction methods will continue to be a primary concern.

The monitored type of performance standard is still in its infancy. The tentative efforts in this direction tend to be in the form of loose guidelines rather than precise specifications and directed towards building owners and management. Future concerns are likely to focus on methods for obtaining a better understanding of existing energy use patterns and on forging a better link between the two types of performance standard (prediction and monitored).

Some of those involved in formulating building energy performance standards have advocated the use of weighting factors to balance the primary resources involved in the extraction, manufacture, and distribution of the various fuels. This is a difficult exercise but one that we believe is best dealt with at the national level if building owners and designers are to make decisions that are consistent with national concerns.

V. OWNER CONCERNS

As described in Chapter 4, there is increasing evidence that building owners are becoming interested in the energy performance of their buildings. Whether owners finance, develop, build, invest in, lease, or occupy a building, they are unlikely to practice energy conservation without some incentive. The incentive for owner interest has been an increasing awareness that operating costs are rising, principally because of rising energy costs.

While it is tempting to see this price mechanism as the principal method of improving energy efficiency, we believe that rational decisions will only be made when the costs in energy terms are understood too. Costs and benefits can be balanced in energy terms; in money terms imbalances may occur which are irrelevant to the energy efficiency of the individual building. That is not to say that the money cost is unimportant; at the very least it can provide a method of ranking potential actions in relation to their importance to the firm. The person who has the responsibility for making decisions which affect the energy use of a building must be aware of the consequences in energy terms. There remains much to be done at a research level before this goal is attained. Before building owners understand the energy consequences of their decisions about the location of a building, the activity it houses or its size, two principal fields of research must be pursued in greater detail.

The first field is the perhaps obvious one of improved, more detailed, multifactor empirical observation and analysis. Much of the analysis that is presented in Chapter 4 has formed the foundation of this field. The multifactor aspects of these studies are especially important. Much of the analysis thus far reported has attempted to examine the influence of each factor in isolation. However, it is unlikely, for example, that a building of a particular size in a particular location (climate), will have the same energy use characteristics as a building of the same size in another location, or even that the relative advantages of alternative buildings will hold in different climates. As well as more detailed research information, it is necessary, therefore, to start to generalize the mutual interactions of the individual factors, especially as they affect building energy performance.

The second field of research is the communication of research results to the people who will be able to use them. It is essential that research data be converted into forms applicable at each level of the exo-endo classification system, so that those people whose decisions may influence the energy performance of buildings are aware of the energy consequences of their decisions. This field of research is of more importance than attempting to improve the level of detail studied. Without this communication, little will be achieved in the quest for improved building energy performance.

It is likely that the presentation of such information in money terms will have the greatest impact. However, as suggested earlier, this approach often introduces too many distortions, of the type discussed in Chapter 4, to be truly effective in improving the energy performance of a building.

Building owners must obtain a firm grasp of the terminology and, more importantly, the quantitative impact of their decision making as it affects energy use in buildings.

VI. DESIGNER CONCERNS

For the building designer, energy is just one of several major concerns, but one that is assuming increasing importance. Much of the emphasis so far has been on operating energy consumption but there is evidence to support the argument that the energy required in the construction of buildings is also an important area for energy conservation.

A. Capital Energy Requirements

As shown in Chapter 5, an informed and rational choice in materials from among the standard ranges can substantially reduce the energy required for construction. However, current design data and methods are inadequate if one wishes to reduce this capital energy requirement.

1. The Limitations of Energy Analysis

Before designers can carry out capital energy requirement evaluations and analyses, we need to establish a readily accessible and accurate data base. We also need widely accepted conventions and methods for carrying out the analyses.

The methods of Energy Analysis are still (see Chapter 5) lacking in several important aspects. First, there is no readily accessible data base that is comprehensive and complete. Except for Stein's data for the U.S., most other energy coefficient data is incomplete for some materials and processes. Second, the available energy coefficient data is accurate to only a first approximation, preventing more accurate analysis from being carried out. Third, the energy coefficients are drawn from input-output analysis, which is usually outdated by about 5 years and often contains insufficient breakdown of information where two or more products are grouped together (since companies producing varied products are classified according to their main products.) Fourth, there is no clear agreement on methods and conventions to be adopted in Energy Analysis, resulting in differences in results simply because of differences in method and depth of analysis. Fifth, and very importantly, the lack of a clear framework and method for integrating human labor and environmental costs into Energy Analysis is a major limitation.

2. Future Developments

Future priorities for Energy Analysis are mainly in the setting up of energy coefficient data banks in a readily accessible form. Further research is also needed in the following areas:

- Analysis of the energy demands of the materials used by the "solar" industry to determine the overall energy benefits for different designs and systems.
- Setting up of a more widely acceptable framework of conventions and analysis methods which integrate human labor and environmental costs into the overall analysis.
- Integration of capital energy analysis into overall design evaluations which would involve calculations of the energy cost in use.

Except for very small buildings it is not always practical to carry out capital energy calculations manually. Thus, the setting up of a readily accessible computer data base should be a high priority. Computerization of the analysis would enable further breakdowns of the energy requirements into fuel types, allow parallel human labor and environmental cost calculations to be carried out more readily, and at the same time yield more useful results. The computer-based energy framework should preferably be compatible with those used by quantity surveyors so that the data would be easily used both for dollar value and for energy requirement estimations.

B. Systems Energy Consumption

This section examines the future concerns of designers, relating mainly to methods of predicting energy use and the assessment of the factors that affect energy use in practice. Although systems energy consumption has been the subject of considerably more work than capital energy requirements, there is still much to be done, as indicated in Chapter 6.

1. Prediction Methods

The need to predict the energy consumption of a building will continue, both for optimal equipment selection and as a vital input to economic analyses. The methods employed to predict the energy consumption of a building are likely to develop in a number of important areas.

The testing of prediction methods will continue to be an area of importance for some time. A better understanding of their accuracy and modeling limitations is needed if these methods are to gain wide acceptance. A number of modeling limitations are already being addressed and some of these, such as the modeling of natural infiltration, may require changes in input data requirements.

As their developers gain a better understanding of the accuracy requirements of individual routines there will be considerable changes in prediction packages. For example, the important effect of artificial lighting on the total energy cost of many buildings must ultimately be reflected in more sophisticated modeling of lighting design and operational factors. Many routines will become simpler and, ultimately, the modeling sophistication of a particular routine should become more directly related to its importance.

Prediction methods are becoming more user friendly, with better default options and more interactive input and output. Packages designed to run on smaller personal computers are already proliferating. Such packages should offer better documentation and be updated more regularly than was the case with the earlier generation of prediction packages. The spread of interactive (often color) graphics to smaller and lower cost computers offers exciting prospects of improved communication for input and output between the program and the program user.

The modeling of control system responses and their effect on the energy-consuming services must be an immediate concern. As their behavior can play a major role in determining building energy consumption, simply assuming that ideal controls are being used is no longer adequate.

Predictions based on actual building energy consumption (empirical data) should become much more common, as the results of more studies of the energy use in large groups of buildings become available and are further analyzed. Such predictions (based on empirical data) would give designers a yardstick against which they could compare the results of detailed predictions for buildings at the design stage. This would provide a useful check for major input data or modeling deficiencies.

Empirical techniques would also allow designers to obtain an initial estimate of the relative importance of building factors affecting systems energy consumption. For example, if the expected lighting electricity cost is high then the resources devoted to energy-conscious lighting design should also be high.

2. Energy Use in Practice

As further results and analyses of studies of actual energy use in buildings become available, the relative influence of the various factors that affect energy consumption should become clearer. For example, the climatic dependence of building energy use bears further study, especially in buildings with high internal gains, where the effects of outside conditions may be small or contrary to that expected. These studies will help quantify the influence of the design of the building fabric.

The actual dependence of energy use on the characteristics of the energy-consuming equipment is an area that merits increasing concern and study. Obtaining better information on the influence of control systems, on the actual energy use behavior of energy-consuming equipment, will be of increasing importance. Controls will be better integrated into the whole system, with higher standards of commissioning, balancing, and maintenance than has frequently been the case in the past.

The behavior of the occupants will increasingly become a designer concern as more people-responsive controls are used. The education of the building occupants in how to best control the building systems for maximum comfort and minimum energy use will also become an area of more concern.

The building designer has to address the present lack of feedback on all aspects of the building performance. The energy consequences of enhanced feedback are manifold and include a comparison of actual energy use with that predicted, evaluation of the thermal comfort conditions actually achieved, identification of areas of potential energy use reduction, and a better idea of how predicted occupant-related factors actually compare with those found in practice.

VII. USER CONCERNS

In Chapter 7 we concentrated on the needs of energy management. Several existing programs were outlined and some tentative conclusions drawn, even though most of the programs were at an early stage in their development. As with so many aspects of building energy performance, detailed feedback on the outcome of such energy management programs is essential if they are to be successful and if useful lessons are to be learned. It is of major concern to us that such feedback takes place.

Four basic energy management techniques were presented in Chapter 7, together with a strategic framework for their use. The four techniques (Pareto analysis, indexes of energy conservation potential, load profiles, and performance lines) can be used both for initiating and for monitoring energy management programs. Each of these is relatively new and further development and testing of each technique is vital to assess their full potential.

Pareto analysis is the logical first step when initiating a conservation program in a large group of buildings. Once the energy consumption of a group of buildings has been significantly altered by a management program, the method can then be used to redefine a target group of buildings. In addition, once the magnitude of energy end uses of each building has been established, Pareto analysis can be applied to any group of building subsystems to determine the specific conservation actions which should be implemented first.

Performance indexes have been developed, for interbuilding comparisons of energy efficiency and conservation potential, based on the summation of all energy uses in each building. There is further scope to refine these indexes, so that they are more precise indicators, by incorporating other significant normalizing factors (e.g., building activity, HVAC equipment type). However, the aggregate nature of energy performance indexes tends to make them more suitable for initiating management programs rather than as detailed monitoring techniques. If building energy consumption can be separated into end uses, then performance indexes for each end use can provide useful monitoring of that function. Even though limitations exist with respect to normalizing factors and responses to climatic variation, there seems little doubt that substantially improved indexes will be developed.

Load profiles and performance lines are essentially monitoring methods for individual buildings. However, their application can be widened to cover other management levels. Monthly load profiles, for each type of fuel used by a group of buildings, can be used to estimate the future energy expenditure of the group. At the systems and components level, load profiles can be used to monitor each item of energy-consuming equipment. The major limitation to its use at this level are existing energy distribution and metering arrangements within buildings which often combine different end uses on the same circuit. However, a range of sensors are available which can overcome at least some of these problems. A major future concern is the extent and frequency of monitoring required for particular energy management purposes.

The last of the four techniques, the performance line, has been developed as a means of establishing a baseline against which actual consumption can be compared. The technique can involve comparison of a single reading with an established performance line, or the comparison of two or more lines each of which comprises a number of readings. While both methods can detect anomalies in energy consumption, the second method has the potential to identify possible causes of such anomalies. Considerable scope exists for the development of this technique as a useful diagnostic tool for energy management.

Of course, energy managers represent only one class of building user. There is a growing awareness that the occupant may have a significant impact on the energy performance of a building and systematic research into this aspect has already begun. It is all too easy to ignore the primary reason for energy use, the needs of the people in the building. This aspect is a vital future concern at the user level in the exo-endo classification system.

VIII. FINAL CONCERNS: BUILDING ENERGY PERFORMANCE

Before summarizing our final concerns it is appropriate to remind readers, and ourselves, that energy performance is just one of a number of factors that must be considered by nations, owners, designers, and users of buildings. Cognizant of this broader context, we have attempted to present the results of our studies in such a way as to assist their integration into the building process as a whole.

Our final concerns are for the further development of methods and feedback mechanisms appropriate to the needs at each level of the exo-endo classification system. We believe this system enables the concerns of those involved in the building process to be clearly identified and provides a framework for new insights. Having complained of the lack of systematic feedback and demonstrated the usefulness of the small amount of empirical data that is available, our final plea for the future is for all those concerned with buildings to study what is happening in them.

We look forward to the time when everyone in the building industry is fully familiarized with the concepts and the range of actual values involved in building energy performance. When, as with vehicle performance, a "good" figure can be readily distinguished from a "bad" one, then the field of building energy performance will have come of age.

CONVERSION FACTORS

$$1m^2 = 10.76 \text{ ft}^2$$
$$1 \text{ MJ} = 948 \text{ Btu}$$
$$1 \text{ MJ} = 10^6 \text{ J}$$
$$1 \text{ GJ} = 10^9 \text{ J}$$
$$1 \text{ TJ} = 10^{12} \text{ J}$$

$$1 \text{ PJ} = 10^{15} \text{ J}$$
$$1 \text{ kWh} = 3.6 \text{ MJ}$$
$$1 \text{ t} = 1000 \text{ kg}$$
$$1 \text{ MJ m}^{-2} \text{ yr}^{-1} = 88 \text{ Btu ft}^{-2} \text{ yr}^{-1}$$
$$1 \text{ MJ kg}^{-1} = 429 \text{ Btu lb}^{-1}$$
$$1 \text{ MJ t}^{-1} \text{ km}^{-1} = 1526 \text{ Btu t}^{-1} \text{ mi}^{-1}$$
$$1 \text{ MJ } \ell^{-1} = 3591 \text{ Btu gal}^{-1} \text{ (U.S. gal)}$$
$$1 \text{ Wm}^{-2} = 0.317 \text{ Btu hr}^{-1} \text{ ft}^{-2}$$
$$1 \text{ Wm}^{-2} {}^{0}\text{C}^{-1} = 0.176 \text{ Btu ft}^{-2} \text{ hr}^{-1} {}^{0}\text{F}^{-1}$$

For the purpose of rough estimates, the Australian, Canadian, New Zealand and U.S. dollar may be assumed to have approximately the same value; and the Pound Sterling (U.K.) to have a value of around $2.00.

INDEX

I

K

L

M

N

O

P

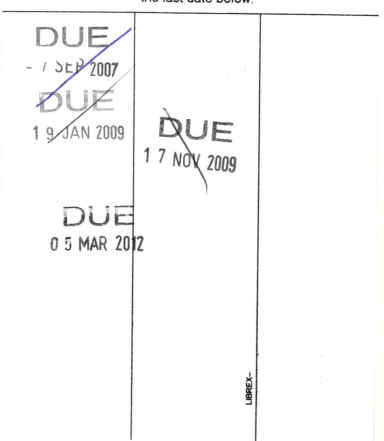